Advance Praise for Television 2.0

"The '2.0' label may have become a buzzword, but *Television 2.0* skillfully puts streaming/downloading hype to the test. Rhiannon Bury draws brilliantly on original empirical data to show how television today remains crucially framed by both domestic and affective relations. Hybridising Deleuzian theory with classic TV studies' work from the likes of Roger Silverstone and James Lull, *Television 2.0* explores the fascinating assemblages and reassemblages of contemporary TV. And Bury makes a vital intervention into debates around fandom and participatory culture by introducing the notion of a 'participatory continuum.' *Television 2.0* is provocative and compelling, well evidenced and astutely argued; I am already a devoted fan of this book."

Matt Hills, author of *Fan Cultures* and co-director of the
Centre for Participatory Culture, University of Huddersfield

Television 2.0

Steve Jones
General Editor

Vol. 102

The Digital Formations series is part of the Peter Lang Media and Communication list.
Every volume is peer reviewed and meets
the highest quality standards for content and production.

PETER LANG
New York • Bern • Frankfurt • Berlin
Brussels • Vienna • Oxford • Warsaw

Rhiannon Bury

Television 2.0

Viewer and Fan Engagement with Digital TV

*John —
Thanks so much for being a part of the study and making this book possible!*

PETER LANG
New York • Bern • Frankfurt • Berlin
Brussels • Vienna • Oxford • Warsaw

Library of Congress Cataloging-in-Publication Data

Names: Bury, Rhiannon, author.
Title: Television 2.0: viewer and fan engagement with digital TV / Rhiannon Bury.
Description: New York: Peter Lang, 2017.
Series: Digital formations, vol. 102 | ISSN 1526-3169
Includes bibliographical references and index.
Identifiers: LCCN 2017019806 | ISBN 978-1-4331-5313-6 (hardback: alk. paper)
ISBN 978-1-4331-3852-2 (paperback: alk. paper) | ISBN 978-1-4331-3870-6 (ebook pdf)
ISBN 978-1-4331-3871-3 (epub) | ISBN 978-1-4331-3872-0 (mobi)
Subjects: LCSH: Television viewers—Social aspects.
Television programs—Social aspects. | Digital television—Social aspects.
Television broadcasting—Technological innovations.
Online social networks—Social aspects. | Fans (Persons)—Social aspects.
Classification: LCC PN1992.55 B87 2017 | DDC 302.23/45—dc23
LC record available at https://lccn.loc.gov/2017019806
DOI 10.3726/978-1-4331-3870-6

Bibliographic information published by **Die Deutsche Nationalbibliothek**.
Die Deutsche Nationalbibliothek lists this publication in the "Deutsche
Nationalbibliografie"; detailed bibliographic data are available
on the Internet at http://dnb.d-nb.de/.

The paper in this book meets the guidelines for permanence and durability
of the Committee on Production Guidelines for Book Longevity
of the Council of Library Resources.

© 2017 Peter Lang Publishing, Inc., New York
29 Broadway, 18th floor, New York, NY 10006
www.peterlang.com

All rights reserved.
Reprint or reproduction, even partially, in all forms such as microfilm,
xerography, microfiche, microcard, and offset strictly prohibited.

Printed in the United States of America

TABLE OF CONTENTS

Acknowledgments	vii
Introduction	1
Chapter 1. Assembling Television: From the Radio to the Internet	15
Chapter 2. Household Assemblers: Patterns of Multiscreen and Multimodal Viewing	37
Chapter 3. Television 2.0 and Everyday Life	55
Chapter 4. Affect and the Television Text	73
Chapter 5. Fandom 2.0: Six Degrees of Participation	91
Conclusion: Rhizomatic for the People	111
Appendix: Television 2.0 Survey Questions	119
References	125
Index	135

ACKNOWLEDGMENTS

I began conceptualizing a research project on "Television 2.0" in 2010 after attending two conferences the previous year on the future of TV—Unthinking Television: Visual Cultures Beyond the Console (George Mason University) and The Ends of Television (University of Amsterdam). The scale of the study was only possible because of the generous research incentive grant awarded to me by Athabasca University. I cannot thank Henry Jenkins enough for offering his support for the project by tweeting the link to the survey and interviewing me about the project on his blog. His generosity enabled me to recruit a larger and more international pool of respondents than would have been possible through my own networks. I would also like to thank the organizers and attendees of the 2010 Flow Conference in Austin, Texas, for taking and/or helping me to promote the survey. The survey and interview data took ten months to collect with the invaluable assistance of three research assistants: Clayton Clemons, who expertly set up and maintained the survey on the university website and exported the raw data into Excel and SPSS; Melanie Cook, who deftly managed the project, most importantly by arranging and keeping track of the interviews; and Fiona MacGregor, who interviewed participants and did some of the data coding in NVivo. Thank you all! I owe a debt of gratitude to Johnson Li, whose expertise in inferential statistical analysis was invaluable, and for coauthoring the journal article published in New Media & Society in 2015 based on this analysis. The people who deserve the most thanks are of course the research participants: the 671 survey respondents who took the time to complete a detailed six-section survey and the 72 interviewees who spent an additional one to three hours on the phone or Skype with Fiona or myself. I would like to single out Kevin Barnhurst, who passed away too young in 2016, for his contributions as a participant and as a scholar.

The road to the completion of this book was a longer one than expected. I wish to thank Mary Savigar at Peter Lang, who approached me about writing a second monograph for the press, and convinced me that I had a book somewhere in all the data. I also appreciated her patience as I missed one contract deadline after another. A shout out to Kathryn Harrison, Sophie Appel and Janell Harris for their help in finalizing and preparing the manuscript for publication as well as to Steve Jones for his constructive feedback on the manuscript. I also want to offer many thanks to Barney Wornoff for kindly offering to apply his creative talents to the design of *Television's 2.0's* fabulous cover.

I am appreciative of the feedback, including that from Lucy Bennett, Paul Booth, Melissa Click, Alice Marwick, and Suzanne Scott on the conference papers, journal articles, and book chapters based on the research project that inform this book. Some of the statistics included in Chapters 1 and 2 also appear in "Is it Live or is it Timeshifted, Streamed or Downloaded? Watching Television in the Era of Multiple Screens" (*New Media & Society*). Versions of the discussion on piracy in Chapter 4 and some of the ideas on fan practices in relation to social media in Chapter 5 appear in "Television Viewing and Fan Practice in an Era of Multiple Screens" (*The Sage Handbook of Social Media*). The idea of the participatory continuum and the discussion on the practices of information seeking and interpretation appear in a similar form in "'We're not There.' Fans, Fan Studies and the Participatory Continuum" (*The Routledge Companion to Media Fandom*). Finally the practice of community making raised in Chapter 5 is discussed more extensively in "Technology, Fandom and Community in the Second Media Age" (*Convergence: The International Journal of Research into New Media Technologies*).

Without the unflagging support of Lee Easton, who spent many an hour on the phone serving as a sounding board as I worked and reworked *Television 2.0*'s central themes and chapters, I never could have completed this book. Thank you dear friend. I also wish to thank those friends and colleagues who lent an ear when the going got tough and kept me going: Alison Chant, Jenny Foreman, Manijeh Mannani, Karen Nelson (riding coach extraordinaire), Meenal Shrivastava, Lorna Stefanick, and Karen Wall. Finally, I extend my love and gratitude to my mother Nancy Bury for always being there for me, and my husband Luis Marmelo, who did all the grocery shopping and cooked every dinner without complaint in that final two-month sprint toward the finish line.

INTRODUCTION

> Sorry I can't come out tonight, I have a date with Netflix and a few bags of Doritos.
> —Esther the Wonder Pig (Jenkins & Walter, 2015 September 18)

Pigs may not yet be able to fly but if the Facebook account for the celebrity porcine is any indication, they are joining the increasing numbers of those who are redefining what it means to watch TV. In recent years, television content has been decoupled not only from the broadcast schedule through the use of digital video recorders but from broadcasting itself through streaming and downloading platforms. Moreover, television content has been decoupled from the television screen itself, the same Web 2.0 technologies enabling viewing on computers, laptops, tablets, and mobile phones. In the early 2000s, television scholars such as Lynn Spigel began to consider the implications for television as a "medium in transition":

> If TV refers to the technologies, industrial formations, government policies, and practices of looking that were associated with the medium in its classical public service and three-network age, it appears that we are now entering a new phase of television—the phase that comes after "TV." (Spigel, 2004, p. 2)

Much of the literature on the changes to television to date has focused on the changes to those industry formations as they relate to production and distribution as well as the efforts by regulatory bodies to respond to such changes (see Bennett, 2008; Bennett & Strange, 2011; Holt & Sanson, 2014; Lotz, 2009, 2014). Empirical study of shifting viewing practices, however, has been left largely to audience measurement and marketing research firms (the Nielsen Company being the juggernaut), government agencies, and independent scholarly organizations such as the Pew Research Center (US). I began the research on which this book is based to bridge the gap not simply between the academy and the television industry but also between fields of study, namely television studies, new media/internet studies, reception studies, and fan studies. At its broadest, this book critically examines what it means to be both a television viewer and a media fan in what Mark Poster (1995) refers to as the second media age. In the following pages, I will discuss the central concepts and themes that provide an analytical framework, describe the Television 2.0 research project, and outline the rest of the chapters.

Television as Assemblage

Until the 1970s, television was generally conceptualized as a form of mass communication or a mass medium, the origins of which date back to theories of mass society that emerged in the early twentieth century:

> Alongside crowds, publics, and social movements, masses are distinguished by their large size, anonymous nature, loose organization, and infrequent interaction. As such, the concept of a mass connotes a group ripe for manipulation and control. (Buechler, 2013)

Concerns about fascist and communist propaganda during the interwar period led to the development of a *direct effects* model of mass communication, described by Elihu Katz as follows:

> There were the mass media, on one hand, sending forth their message, and the atomized mass of individuals, on the other, directly and immediately responding—and nothing in between. (cited in Lubken, 2008, p. 23)

In the same period, Marxist theorists, most famously those associated with the Frankfurt School, were engaged in vigorous critiques of popular culture and the cultural industries that produced it. After taking on film and radio, Max Horkheimer and Theodor Adorno turned their attention to television in the early 1940s, before its widespread adoption:

> Television aims at a synthesis of radio and film, delayed only for so long as the interested parties cannot agree. Such a synthesis, with its unlimited possibilities, promises to intensify the impoverishment of the aesthetic material so radically that the identity of all industrial cultural, still scantily disguised today, will triumph openly tomorrow in a mocking fulfillment of Wagner's dream of the total art work. (2002, p. 97)

Adorno would continue to express grave concern about the "nefarious" effects of television and advocate for a behaviorist approach to better understand the "socio-psychological stimuli of televised material" (1954, p. 213). He was particularly concerned about the ways in which overt and hidden messages operated in tandem to "channelize audience reaction" (p. 222). Although he admitted that it is difficult to "corroborate by exact data," he nonetheless stated that "majority of television shows today aim at producing or at least reproducing the very smugness, intellectual passivity, and gullibility that seem to fit in with totalitarian creeds even if the explicit surface message of the shows may be antitotalitarian" (p. 222). While later iterations of these critiques and approaches were more nuanced, Raymond Williams (1975) argues that the analytic of mass communication serves to rationalize a focus on effects.[1]

Williams was one of the first scholars to untether television from its mass media traditions and define and analyze it foremost as a "technology of social communication" which was "preceded by and continues to overlap with other forms of social communication within social groups and specific institutions" (p. 21). Roger Silverstone pushes the analysis further:

> Seeing technology as a system involves, above all, seeing technology as both a material and social phenomenon. Relations between objects and artifacts; relations between people and institutions; the power of the state and politics of organizations; the embeddedness of the systemic relations of technology in a constantly vulnerable environment of social, political and economic structures: all of these elements define a framework from which new technologies emerge, old technologies are discarded, and from which all technologies are produced and consumed. (1994, p. 84)

Silverstone defines television as a *tele-technological system* that is distinguished from other communications technologies by its articulation into the household, as both an object and as a broadcast medium. While of course television can be viewed in a number of public and private spaces (see McCarthy, 2001), the home is the primary site of reception in Western societies. The purchase of TV sets and, more recently, computers and mobile devices, along with the devices and services associated with delivery (e.g., cable) ties users into relations of consumption. At the same time users cannot be reduced to passive consumers; rather, they actively integrate these televisual technologies into the various spaces of the household for individual and collective use. "Through the structure and contents of its programming," Silverstone contends, television as a broadcast medium "draws members of the household into a world of public and shared meanings as well as providing some of the raw material for the forging of their own private, domestic culture" (p. 83).

Silverstone recognizes that the household is a site of struggle over the production of meaning, a site in which "the certainties of domination become the uncertainties of resistance" (p. 79). John Fiske makes a similar point: "In going out to cinema, we tend to submit to its terms, to become subject to its discourse, but television comes to us, enters our cultural space, and becomes subject to our discourses" (1987, p. 74). The television text is polysemic, argues Fiske, "a potential of unequal meanings, some of which are preferred more, or proffered more strongly, than others" (p. 65). As such it "offers provocative spaces within which the viewer can use her or his already developed competencies" (p. 95). This understanding of signification is at the heart of the "reception turn" in media and communication studies and casts into sharp relief the limitations of the transmission model of mass communication. Although the negotiation of meaning between text and viewer will not be analyzed specifically in this book, it is understood to be the process that underpins all engagement with television texts and the practices they engender.

Engagement with television, however, is not just about the fashioning of a private domestic culture; it also can involve the fashioning of a participatory fan culture, or *fandom*. Henry Jenkins (1992) was one of the first scholars to distinguish between bystanders (casual viewers) and fans. Based on his research with fans of series such as *Star Trek*, *Beauty and the Beast*, and *Twin Peaks*, he argues that fan engagement is "a process, a movement from the initial reception of a broadcast toward the gradual elaboration of the episodes and their remaking in alternative terms" (p. 53). He challenges the common perception of fans as obsessive and incapable of critical judgment due to their intense emotional connection to a series, writer, actor, etc. Fans, he argues, are *textual poachers* who do not passively accept the meanings offered by content producers. Instead they engage in a range of interpretative and creative practices, including discussing the series with other fans; attending fan conventions; producing fan fiction, art, music, and video; and engaging in fan activism. Jenkins laid the groundwork for the development of the field of fan studies, the result of which was a divergence from reception studies and its central focus on television as a domestic medium.

To capture the complexities of viewer engagement with television texts in today's era of multiple screens, content sharing, and social networking, it is imperative to recognize that such engagement is shaped by both domestic and affective relations. Engaging with television as a member of a household is bound up with the everyday (Lull, 1990; Silverstone, 1994), that is, the routines of work and leisure of all members of the household. The *modes of reception*, from background/distracted viewing to attentive and focused viewing, will depend on how television is "inserted" into particular moments of one's domestic routine (Fiske, 1987, p. 146). While one may not always care about what is on TV just as long as it is on, one is also unlikely to attentively view the same programming week after week without some kind of emotional investment or pay off. Drawing on Roland Barthes' notion of the pleasure of the text, Fiske argues that television texts afford many pleasures to viewers because of their polysemic construction, which enable viewers to match the discourses circulating in the text with their own subjectivities. This notion, however, falls short of explaining the emotional attachment of fans to specific texts. Larry Grossberg's (1992) conceptualization of *affect* is useful in this regard. Distinguishing it from both meaning and pleasure while not reducing it to the level of the personal or purely subjective, he defines it as a feeling or sensibility that gives "'color,' 'tone,' or 'texture'" to the experiences of everyday life (p. 585). Affect is what determines the strength of investment in the meanings and pleasures of particular genres and texts. This affective relationship to the television text creates the conditions for committed, fannish viewing. I take a similar position to that of Jenkins (1992) in that I do not consider viewing in itself to be a form of participation; rather, it underpins all participatory fan practices and creates potential linkages between domestic culture and participatory culture.

In light of the above discussion, I find it more fruitful to refer to television as a *tele-technological assemblage* rather than a system. Postmodern theorists Gilles Deleuze and Felix Guattari use the term to emphasize the instability and temporality that underlie the "holding together of heterogeneous elements" as well as processes of "becoming and unbecoming, combining and recombining" (1987, p. 323). This concept is useful to social science researchers because it addresses the problem of heterogeneity while "preserving some concept of the structural" (Marcus and Saka, 2006, p. 102). The process of assemblage is not linear but more akin to that of a biological rhizome, extending in ways that are not always intended or predictable or containable. Thus through fan engagement, television's meanings can be extended beyond the site of reception into the spaces of participatory culture. I would add that coherence, or lack thereof, across the nodes of the assemblage is uneven. The highest degree of stability is located in the institutions that comprise the television industry—the production companies, the networks, and the national regulatory bodies. Conversely, the lowest degree of stability is located at the sites of reception and participation, where meanings, pleasures, and affective relations are made and remade.

In light of the above, I argue that television is not simply a medium in transition as Spigel (2004) suggested; rather, its institutions, texts, and viewing and participatory practices are in the process of *reassemblage*, more specifically as a result of hybridization with the internet. Poster was one of the first scholars to pinpoint the role of information and communication technologies (ICTs) in starting such a process:

> With the incipient introduction of the information "superhighway" and the integration of satellite technology with television, computers and telephone, an alternative to the broadcast model, with its severe technical constraints, will very likely enable a system of multiple producers/distributors/consumers, an entirely new configuration of communication relations in which the boundaries between those terms collapse. A second age of mass media is on the horizon. (1995, p. 2)

From the vantage of the present it is clear that these boundaries have yet to collapse; that said, I will argue that television–internet hybridity challenges the linearity and centralization of the first media age. As Lisa Parks puts it, "the historical practices associated with over-the-air, cable and satellite television have been combined with computer technologies to reconfigure the meanings and practices of television" (2004, p. 134). The title of my research project and this book, Television 2.0, was chosen to emphasize the continued legibility of television but also its hybridity. Web 2.0 was coined by Tim O'Reilly in 2004 to envision a new approach to dotcom commerce that relied on direct consumer collaboration and engagement. Although it has been taken up broadly, reduced to an empty buzzword in many contexts, it remains a useful label with which to describe a cluster of technologies and platforms that are distinct from previous internet technologies. According to Henry Jenkins, Sam Ford, and Joshua Green (2013), "the mechanisms of Web 2.0 provide the preconditions for spreadable media" (p. 49). They define spreadability as "the continuous process of repurposing and recirculating" of mass content by individuals and, communities (p. 27). They make a similar argument to that of Poster about internet technologies, namely that spreadability breaks down "the perceived divides between production and consumption" (p. 27). They contrast it to both the traditional media industry notion of "stickiness" and the new media notion of user-produced content, or *produsage* (Bruns, 2006). Instead they talk about "user-circulated content," a description which recognizes the continued significance of mass media content but recognizes the role of streaming and downloading platforms in spreading content outside the national broadcasting context of established global flows. It is these same mechanisms that have enabled the development of social networking sites such as Facebook and Twitter. Collectively known as social media, these platforms provide new possibilities of fan participation and engagement.

Researching Television 2.0

The TV 2.0 research project was designed to fill a gap in the literature by providing a theoretically informed, empirical study of television's reassemblage in relation to reception and participatory practices. It is primarily a qualitative study with a quantitative component. To collect the data I used an online survey questionnaire and telephone/ Skype interviews. The questionnaire was created using LimeSurvey open-source software and hosted on the Athabasca University server. The survey was divided into six sections of predominately closed questions: Section A asked for demographic information, the results of which are presented below. Sections B, C, and D asked a series of questions about the viewing of television programming on television, computer, and mobile screens, respectively. Section E asked about genres of programming (e.g., drama, comedy, news, and sports) in relation to the platforms used to watch them (broadcast, internet, and DVD).[2] Section F was the longest section and was designed to collect data on fan practices and involvement in participatory culture. Although one can be a fan of a range of genres, the questions focused on television's dominant form, the serial narrative (Fiske, 1987), as this is the form around which participatory culture has formed. The section began with an exclusion question that asked the respondents to define themselves as fans. Those who did not engage in at least one of the practices listed below the question were taken to the end of the survey.

The survey was piloted in June 2010 and went live in September 2010 during the biennial *Flow* television studies conference, held at the University of Austin, Texas. Purposive and snowball sampling were used to identify a reasonably diverse English-speaking (first or second language) television-viewing population in general and a television-fan population in particular. Individuals and online communities with scholarly, professional, and/or personal interests in television, media studies, and/or fandom were invited to take the survey via email, listservs, and social media (Facebook and Twitter) and encouraged to pass on the link to and information about the survey to their personal and professional networks. Thus the population was not expected to be representative of a global television-viewing population.

The survey closed in April 2011 with a total of 998 surveys attempted and 671 completed. Just over forty percent of the respondents agreed to be contacted for an interview. Of those 281, 110 reconfirmed their interest in taking part. The interviews began in April 2011. The first fifty participants were selected in the order in which they responded; purposive sampling was used after a review of the demographic data to select the final set of participants to ensure a more diverse sample. A preliminary review of the data suggested that saturation had been reached at seventy, already a large data set given the resources of the project. A total of seventy-two semi-structured interviews were completed by September 2011, ranging from thirty minutes to two and a half hours. The questions were customized for each participant based on their survey responses. For those who indicated involvement in participatory culture, a topical life history was built. Sandra Kirby, Lorraine Greaves and Colleen Reid describe such a history as "similar to a life history except that only one part of a person's experience is described" (2006, p. 160). Follow-up email exchanges with select participants for clarifications and additional detail were completed as required.

SPSS software was used to analyze the survey data. I worked with a statistical consultant, Johnson Li, to produce descriptive and inferential analyses. In terms of the latter, we ran regression analysis to predict increases or decreases with "live" viewing in relation to newer viewing modes such as time-shifted, online, and mobile. We also ran an analysis of variance (ANOVA) to assess if any subgroup differences existed among demographic variables of gender, age, and region of residence. QSR NVivo software was used to code the interview data and aid in the qualitative analysis. Quotations from the participants are presented at length in Chapters 2–5 in recognition of the value of thick, descriptive data (Geertz, 1973). As Monica Gallant points out, "the value in stories about particular people in a specific context is especially useful ... where the body of published research is limited" (2008, p. 247). As a feminist scholar, I argue that it is also important for the "voices" of the participants to be heard and be recognized as coproducers of knowledge (Kirby, Greaves, & Reid, 2006). A number of participants were also media scholars and/or fans with long histories of involvement in fandom. For example, I learned a great deal about the broadcast contexts outside of Canada and the United States through the interviews. Giving voice does not in itself equalize power relations in the research context; neither does it offer unmediated accounts of experience. This book, like any publication based on qualitative research, is "overinvested in second-hand memories" to quote Deborah Britzman (1995, p. 153). Experiences are reconstructed by the participants themselves and then by the researcher, who codes, selects, and organizes them into a coherent text. To protect confidentiality, the interview participants are identified in the following chapters by a name of their choosing. A few chose to use their social media or fandom identities. In cases where more than one participant chose the same first name, a last initial has been added.

Demographic Snapshot

> Although most of us talk about "television" without any qualifying prefix such as "Australian" or "American," the fact is that, especially since the digital revolution and notwithstanding processes of globalization, "television" involves such varying forms, platforms and content in its different national and regional locations that it is increasingly implausible for one set of experiences to be representative.
> — Graeme Turner (2011, p. 32)

Following Turner, I have tried to "dehomogenize" television by foregrounding not only national and regional variations of viewing but age and gender variations as well. Starting with gender, 445 of the survey respondents identified as female (66.3 percent), 217 as male (32.3 percent), and 9 (1.3 percent) reported their gender as non-binary. Three-quarters of the interview participants were female (fifty-three compared to nineteen males). The age of the survey respondents ranged from eighteen (the minimum age to take part in the survey) to seventy-five, with a mean of 34.6. The age variable was recoded into age groups, with just over seventy percent of the respondents under the age of forty: 39.5 percent ($n = 265$) were in the eighteen to twenty-nine cohort; 30.7 percent ($n = 206$) were in the thirty to thirty-nine cohort; 16.4 percent ($n = 110$) were in the forty to forty-nine cohort; 10.4 percent ($n = 70$) were in the fifty to fifty-nine cohort; and three percent ($n = 20$) were sixty or older. The interview participants were slightly older with just under two-thirds below forty: twenty-four (eighteen to twenty-nine), twenty-two (thirty to thirty-nine), eighteen (forty to forty-nine), six (fifty to fifty-nine), and two (sixty plus).

As for country of residence, the survey gave the respondents three choices: Canada (my country of residence) ($n = 120$); United States ($n = 268$); or "Other," which required them to provide the name of the country. As expected, those residing in North America (57.8 percent) made up the majority, but the sample was more international than I had anticipated: thirty-one other countries were represented. The United Kingdom ($n = 81$; 12.8 percent) was also coded as an individual country because of the number of respondents. The remaining countries were recoded into seven regions modeled on those used by television scholars. Only two of these regions, however, had enough respondents to produce reliable regression and ANOVA results: Europe ($n = 135$) and Australia/New Zealand ($n = 27$). Africa, Asia, South America, South Asia, and Western Asia were recoded as missing values for the reason that there is no common broadcast model or model of service provision across these regions to justify combining them as a single category of "other regions." As for the interview participants, just under half resided in the United States (thirty-three), twelve in Canada, eleven in the United Kingdom, six in Europe (Belgium, Germany, Norway, Netherlands (two), and Serbia), three in Australia, two in New Zealand, and one each in Argentina, Brazil, India, Israel, and Malawi.

Finally, the survey included an optional question in which respondents could self-identify as to their race and/or ethnicity. A review of the completed responses indicated that less than one percent could be coded in a category other than white. I recognize not being able to critically analyze issues of race, reception, and fannish participation is a limitation to this study; at the same time, I want to acknowledge whiteness as a subject location of unearned privilege (Frankenberg, 1993; McIntosh, 1993).

The Rest of the Book

The major themes explored in *Television 2.0* are organized into five chapters, listed and described below.

Chapter 1: Assembling Television: From the Radio to the Internet

This chapter details the processes of assemblage, reassemblage, and hybridization of television, with particular attention paid to viewing and participatory practices. It begins with a discussion of radio and the formation of the public and commercial national broadcast models in the early twentieth century, the period during which radio became a domestic technology. It then examines television's takeover of the radio assemblage and its technological and social transition from the classic network era to the multichannel universe. The last section of the chapter examines the process of digitalization in terms of convergence and divergence, which led to the creation of internet protocol TV (IPTV). Rather than the latter replacing broadcast TV (BTV), IPTV and BTV coexist as separate but overlapping household *intra-assemblages*.

Chapter 2: Household Assemblers: Patterns of Multiscreen and Multimodal Viewing

In this chapter, I look more closely at household engagement with the BTV and IPTV intra-assemblages. The first section draws on the TV 2.0 survey data to trace out the broader patterns of screen and device use as well as modes of viewing—live viewing, time-shifting, and online viewing afforded by streaming and downloading technologies. The data clearly indicate an almost ubiquitous engagement with both intra-assemblages, although there are statistically significant differences across the demographic variables. Based on a fine-grained analysis of the interview data, I argue that viewers can be divided into three categories: those who have strong investments in the BTV assemblage, those who have strong investments in the IPTV assemblage, and those who have no allegiance to either, engaging regularly with both as hybrid TV (HTV) assemblers. A number of data samples are provided to capture the nuances of this engagement within and across these categories.

Chapter 3: Television 2.0 and Everyday Life

Drawing primarily on the literature that conceptualizes television as a domestic technology, this chapter examines both the environmental and social uses of television as a part of everyday life (Lull, 1990). The first half of the chapter will focus on environmental uses, specifically, on background and/or distracted viewing. It also looks at the ways in which television is integrated into daily routines as a leisure activity and as a form of relaxation. The second half focuses on viewing as a social activity with other members of the household as well as with friends, even if not together in the same room. Based on the analysis of the interview data, I argue that multiscreen and multimodal viewing have altered but not fundamentally changed television's imbrication with domestic relations.

Chapter 4: Affect and the Television Text

This chapter first takes a closer look at committed fannish viewing and the affective relationships that are formed around the series of which one is a fan. The survey results demonstrate that the vast majority of viewers are also fans. The interview data allow for a more fine-grained analysis of the ways in which affective intensities are related to investments in particular genres and bourgeois aesthetics, investments that may wax and wane over time. I argue that affective intensity can also determine the choice of viewing mode: fans anticipating new episodes will either watch the original broadcast live or download an unauthorized copy depending on which mode gives them the most timely access. I will also discuss the technologies used to discover and catch up with new series, even those no longer in production. Repeat and marathon ("binge") viewing as well as the practice of DVD collecting are also discussed.

Chapter 5: Fandom 2.0: Six Degrees of Participation

In this final chapter, I discuss four clusters of online fan practices; I have placed these on a participatory continuum, with those requiring the least amount of involvement with fan communities and culture at one end and those requiring the most at the other. I begin with information seeking and then examine collective reaction/interpretation through "lurking" or occasional posting on online forums. Next I turn to community making, which Jenkins (1992) has argued is at the heart of participatory culture. Finally I discuss the production of creative works (fan fiction and fan videos), a practice closely bound up with membership in online communities. I also pay particular attention to the role of social media in altering these established practices and in creating new ones. I conclude that platforms such as Facebook, Twitter, and YouTube disturb the boundaries between viewing, information seeking, interactivity, and community.

Notes

1. One exception was the uses and gratifications approach. At its broadest, it "asks not what the media do to people, but what people do with the media" (Katz, cited in Lull, 1990, p. 29).
2. Close analysis of Section E revealed that the findings were too general to provide useful insights toward a better understanding of changing patterns of reception.

References

Adorno, T. W. (1954). How to look at television. *Quarterly of Film, Radio and Television, 8*(3), 213–235.

Bennett, J. (2008). Television studies goes digital. *Cinema Journal, 47*(3), 158–165.

Bennett, J., & Strange, N. (Eds.). (2011). *Television as digital media.* Durham, NC: Duke University Press.

Britzman, D. (1995). Beyond innocent readings: Educational ethnography as a crisis of representation. In W. Pink & G. Noblit (Eds.), *Continuity and contradiction: The futures of the sociology of education* (pp. 133–156). Cresskill, NJ: Hampton Press.

Bruns, A. (2006). Towards produsage: Futures for user-led content production. In F. Sudweeks, H. Hrachovec & C. Ess (Eds.), *Cultural Attitudes towards Communication and Technology* (pp. 275–284). Tartu, Estonia.

Buechler, S. M. (2013). Mass society theory. In D. A. Snow, D. della Porta, B. Klandermans, & D. McAdam (Eds.), *The Wiley-Blackwell encyclopedia of social and political movements.* Hoboken, NJ: John Wiley & Sons.

Deleuze, G., & Guattari, F. (1987). *A thousand plateaus: Capitalism and schizophrenia* (B. Massumi, Trans.). Minneapolis: University of Minnesota Press.

Fiske, J. (1987). *Television culture.* New York: Methuen.

Frankenberg, R. (1993). *White women, race matters: The social construction of whiteness.* Minneapolis: University of Minnesota Press.

Gallant, M. (2008). Using an ethnographic case study approach to identify socio-cultural discourse: A feminist post-structural view. *Education, Business and Society: Contemporary Middle Eastern issues, 1*(4), 244–254.

Geertz, C. (1973). Thick description: Toward an interpretive theory of culture. In: *The interpretation of cultures: Selected essays* (pp. 3–30). New York: Basic Books.

Grossberg, L. (1992). Is there a fan in the house?: The affective sensibility of fandom. In L. A. Lewis (Ed.), *The adoring audience: Fan culture and popular media* (pp. 581–590). London; New York: Routledge.

Holt, J., & Sanson, K. (Eds.). (2014). *Connected viewing: Selling, streaming, & sharing in the digital era.* New York: Routledge.

Horkheimer, M., & Adorno, T. W. (2002). The culture industry: Entertainment as mass deception (E. Jephcott, Trans.). In G. S. Noerr (Ed.), *Dialetic of enlightenment: Philosophical fragments* (pp. 94–136). Stanford, CA: Stanford University Press.

Jenkins, H. (1992). *Textual poachers: Television fans & participatory culture.* New York: Routledge.

Jenkins, H., Ford, S., & Green, J. (2013). *Spreadable Media: Creating value and meaning in a networked culture.* New York: New York University Press.

Jenkins, S., & Walter, D. (2015). Esther the Wonder Pig. Facebook status update. Retrieved from https://www.facebook.com/estherthewonderpig/

Kirby, S. L., Greaves, L., & Reid, C. (2006). *Experience research social change: Methods beyond the mainstream* (2nd ed.). Toronto: Broadview Press.

Lotz, A. (2009). What is U.S. television now? *The Annals of the American Academy of Political and Social Science, 625*(September), 49–59. doi: 10.1177/0002716209338366

Lotz, A. (2014). *The television will be revolutionized* (2nd ed.). New York: New York University Press.

Lubken, D. (2008). Remembering the straw man: The travels and adventures of hypodermic. In D. W. Park & J. Pooley (Eds.), *The history of media and communication research: Contested memories* (pp. 19–42). New York: Peter Lang Publishing.

Lull, J. (1990). *Inside family viewing: Ethnographic research on television's audiences.* New York: Routledge.

Marcus, G. E., & Saka, E. (2006). Assemblage. *Theory, Culture & Society, 23*(2–3), 100–109.

McCarthy, A. (2001). *Ambient television: Visual culture and public space.* Durham, NC: Duke University Press.

McIntosh, P. (1993). White privilege and male privilege. In A. Minas (Ed.), *Gender basics: Feminist perspectives on women and men* (pp. 30–38). Belmont, CA: Wadsworth.

Parks, L. (2004). Flexible microcasting: Gender, generation and television-internet convergence. In L. Spigel & J. Olsson (Eds.), *Television after TV: Essays on a medium in transition* (pp. 133–156). Durham, NC: Duke University Press.

Poster, M. (1995). *The second media age.* Cambridge, MA: Polity Press.

Silverstone, R. (1994). *Television and everyday life.* New York: Routledge.

Spigel, L. (2004). Introduction. In L. Spigel & J. Olsson (Eds.), *Television after TV: Essays on a medium in transition* (pp. 1–34). Durham, NC: Duke University Press.

Turner, G. (2011). Convergence and divergence: The international experience of digital television. In J. Bennett & N. Strange (Eds.), *Television as digital media* (pp. 31–51). Durham, NC: Duke University Press.

Williams, R. (1975). *Television: Technology and cultural form.* New York: Schocken Books.

· 1 ·
ASSEMBLING TELEVISION
From the Radio to the Internet

Borrowing from Raymond Williams, a tele-technological assemblage takes form through a process in which "scattered technological devices" become "an applied technology and then a social technology" (1975, p. 24). Although such devices did come together to create a viable visual broadcasting technology, television skipped a number of steps, instead taking over the existing assemblage of radio. I begin with a discussion of the domestication of radio, and then trace out the assemblage and reassemblage of analog and digital television, with attention paid to linkages to participatory culture. I argue that the process of digitalization is not linear; rather broadcast TV (BTV) and internet protocol TV (IPTV) have become overlapping household intra-assemblages. This chapter is not intended to provide a comprehensive history; rather, it pays attention to technological and industry developments as they inform reception and participatory practices in the US and UK contexts, contexts which are representative of the commercial and public models of broadcasting in Western democracies.

Radio Days

Radio or "the wireless," as it was first known, has its roots in late nineteenth-century wireless telegraphy and telephony, and was the domain of hobbyists and ham operators until the early 1920s. By then it had coalesced into the familiar model described by Williams as "centralized transmission" and "privatized reception" (1975, p. 30). Unlike cinema for which distribution was controlled by the content producers, sound broadcasting was developed without any specific content (*Ibid.*). "Listening in" to broadcasts of music, lectures, and sermons, produced by anyone who had access to and could operate the equipment, quickly captured the public imagination (Carlat, 1998). "Nothing was fixed," notes Susan Douglas in reference to the American context:

> Not the frequencies of stations ..., not the method of financial support, not government regulations, and not the design or domestic location of the radio itself. There were no networks—known in the late 1920s as the chains—and there was very little advertising on the air. Department stores, newspapers, the manufacturers of radio equipment, colleges and universities, labor unions, socialists, and ham operators all joined the rush to start stations. (1999, p. 56)

Driven by economic and political imperatives, production was soon concentrated and controlled by the state and/or large corporations depending on the national context. In the Western European countries, including the United Kingdom, production and transmission costs were either financed wholly or in part by the state. In the United Kingdom, for example, commercial broadcasting was started by Marconi Wireless in 1920 but banned two years later. The British Broadcasting Company, a consortium of manufacturers, was set up by the government to both sell wireless sets and produce programming that was to be funded through royalties from both sales of sets and a set licensing fee, which was established in 1923. By the end of the year, 200,000 licenses had been issued (Press Association, 2005). In 1926, after a government review, the Company was dissolved and overhauled to become the British Broadcasting Corporation (BBC), a fully public broadcasting service. Jostein Gripsud makes the case that the agricultural origin of the term broadcasting— "the sowing of seeds in as wide (half circles) as possible"—served as "an optimistic, modernist metaphor" that fit with the vision of John Reith, the first Director-General of the BBC, to provide programming that educated, informed, and entertained its citizens (2004, p. 211). In contrast, commercial radio dominated in the United States and in Canada. As early as 1916, David Sarnoff, who later became Chairman of the Radio Corporation of America (RCA), told senior executives at Marconi that enormous profits would be generated if the radio moved out from basements and garages, the domain of hobbyists, and into the living rooms of American families as a central source of news and entertainment (Matelski, 1995). In 1919, after negotiations with the US military and in cooperation with Westinghouse Electric and American Telephone and Telegraph (AT&T), General Electric bought out several wireless manufacturers, including the well-known American Marconi Company, and created RCA. This acquisition gave RCA a monopoly on the retail and marketing of radio sets in America. Although RCA was not involved in broadcasting at that time, other big corporations owned radio stations: the first station to be granted a commercial broadcast license in 1920 was owned by Westinghouse. The first ad was run in 1922 on one of AT&T's stations (Douglas, 1999). These early stations were set up primarily as a means to sell more sets; advertising revenue was secondary, a way to pay for the programming needed to attract more consumers and listeners. Dozens of other stations opened over the course of two years in major American and Canadian cities. In 1926, RCA purchased a core group of stations or "chains" from AT&T and formed the National Broadcasting Corporation (NBC). This expansion was followed by regulation in the form of the Federal Radio Commission in 1927 (becoming the Federal Communications Commission (FCC) in 1934). The Columbia Broadcasting System (CBS) was bought and consolidated by William Paley in 1928 with sixteen affiliates. It had forty-seven a year later (Bergreen, 1980).

According to Silverstone, a technology becomes domesticated when "technological artefacts and delivery systems" are appropriated, controlled, and rendered "more or less invisible within the daily routines of daily life" (1994, p. 98). Before radio could be fully domesticated, thus generating the kinds of profits envisioned by Sarnoff, the tuning technology had to be improved. By the early 1920s, the crystal sets that required individual headsets had been replaced with vacuum tubes which enabled amplification sufficient to project the sound to fill an entire room (Carlat, 1998). In the United States, sales of sets jumped from 60 million in 1922 to 358 million in 1924 (Matelski, 1995). Even so, the set still required an operator, not just a listener. In fitting with normative domestic relations, the operator was assumed to be the male head of the household. As reported in an issue of *Radio World*, tuning required "systematic, scientific manipulation of variable conditions. ... The man who just turns the dial in a childish and unknowing fashion waiting for the 'magic box' to spring a 'hocus pocus' trick is ... cutting himself off from a lot of entertainment" (cited in Carlat, 1998, p. 123). In order to be integrated into household routines for all members of the family, particularly those of women, manufacturers had to effectively "feminize" radio. Mohawk Electric came up with a single-tuning set in 1925 and by 1928, it had become the industry standard (*Ibid.*). Moreover, the set itself was feminized—the more expensive models were consoles that also functioned as furnishing.[1] In 1930, the US Census included a question on radio ownership, finding that thirty-nine percent of households had a set. By 1940, seventy-three percent of all American households had a radio (US Census Bureau, 1999). In the United Kingdom, 12 million licenses have been issued by 1930. By 1939, three-quarters of all British households had a set license (Tomlinson, 1990).

The domestication of the radio was furthered with the establishment of a regular programming day that was between twelve and eighteen hours long. According to a 1938 FCC survey, over half the programming was devoted to music, both live and recorded, with the rest divided evenly among talk, news, variety, and drama (Sterling & Kittross, 2002). The latter category included so-called "prestige" drama anthologies and plays as well as serials and series (*Ibid.*). Although the serial has precedent in popular genres of nineteenth- and twentieth-century fiction, which included detective fiction and westerns (Williams, 1975), it was one of the first forms of entertainment programming produced specifically for radio rather than adapted from vaudeville or theatrical performances. It challenged the programming convention at the time that storylines needed to be resolved by the end of the episode (Matelski, 1995). The serial's history is also intertwined with that of audience measurement. The first US audience survey for radio was conducted by Archibald Crossley and his company in the late 1920s, using another domestic technology—the telephone (Webster & Phalen, 2009). His company went out of business but audience measurement was taken up by none other than Arthur C. Nielsen in 1936 (*Ibid.*). One of the first American serials was *Amos 'n' Andy*. Within a year of its debut in 1928, it was a huge hit and at the peak of its thirty-two-year run, it had an audience of 40 million, one-third of the American population (Nachman, 1998).[2] Michelle Hilmes (1993) argues the series directly influenced the future development of both soaps, which debuted in 1935, and situation comedies (sitcoms). Its popularity also initiated the familiar network practice of the repeat broadcast to ensure east and west coast listeners could hear the program at a convenient time.

In the postwar period, the evening schedule of the American networks was dominated by popular series. Crime thrillers and sitcoms, in particular, made up the largest segment, while the ten am to five pm time slots were dominated by soaps (Sterling & Kittross, 2002). In contrast, the BBC schedules were almost devoid of popular content until 1946 when the Corporation finally began to offer a "Light Programme" of music, serials, and episodic series to complement its prestige productions (Bouckley, 2016). This was a move made in part to compete with Radio Luxembourg, which set out to capture a British audience interested in less highbrow fare, programming that included imported US serials and series (208 Radio Luxembourg, 2001).

Given the focus on effects by early communications scholars, we know very little about reception and participatory practices from the days of radio. There are a few anecdotes, however, that suggest that *Amos 'n' Andy* was "appointment/must-hear" radio at the height of its popularity. According to Marilyn Matelski, when daylight savings time was adopted, "factories changed their hours so employees could get home in time for the show" (1995, p. 9). Similarly Gerald Nachman claims that President Coolidge "refused to be disturbed" during the broadcast and that "water flow dropped dramatically ... and phone lines went still every night at 7" (1998, p. 273). Moreover, *Amos 'n' Andy* inspired what might well be the first instance of media fan activism: NBC received 18,000 letters threatening to boycott Pepsodent, the show's sponsor, if Amo's wife Ruby was "killed off" (Nachman, 1998). Whether these anecdotes are true or not does not matter: there can be no doubt that these committed listeners and letter writers were fans, even if they were only represented as part of a mass audience.

A Set with a View

According to Williams (1975), the transmission of still and moving images was being pursued in Britain, Western Europe, the Soviet Union, and the United States in the first decade of the twentieth century. However, it was not until the late 1920s that functional systems were developed by public and private corporations such as AT&T (in its Bell Laboratories), RCA, the BBC, and Marconi-EMI (UK). The BBC began the first television service in 1936. In the United States, among the first commercial stations on the air were those owned by NBC and CBS (ABC followed in 1948). By 1948 almost seventy companies were manufacturing TV sets in the United States, with the market dominated by RCA, DuMont, and Philco (Boddy, 1990). To promote sales, the BBC and the US networks directed radio's financial and creative resources into producing and broadcasting television programming, both adapting and creating original content. Given the much higher production costs, it is not surprising that these networks were interested in selling not only to national advertisers but to international markets. According to William Marling

(2006), CBS set up the first subsidiary for foreign distribution and by 1958 had made over $4 million from sales of *Lassie*. By 1961, the network was selling 1500 thirty-minute episodes to fifty-five countries (*Ibid.*).

Because visual broadcasting was developed within the existing sound broadcasting model, the TV set arrived in the living room already domesticated. In 1950, less than ten percent of American households owned a television, compared to ninety-one percent who owned a radio (US Census Bureau, 1999). Five years later, the ownership had jumped to sixty-three percent (*Ibid.*). There can be no question that the postwar economic boom and consumer affluence were adoption drivers. Just as he had predicted the commercial success of radio, Sarnoff foresaw that postwar America's new automobile suburbs would be a good fit for TV (Spigel, 1992). A 1953 *Harper's* article describes the suburban landscape as one of "endless picket fences of telephone poles and television aerials" (cited in Spigel, 1992). Spigel argues that the homes were

> strung together like Christmas tree lights on a tract with one central switch. And that central switch was the growing communications complex through which people could keep their distance from the world buy at the same time imagine their domestic spheres were connected to wider social fabric. (p. 6)

Thus the postwar suburban home needs to be understood as a sphere of increasingly privatized consumption and leisure, although as Spigel points out, it was also a site of unpaid labor for wives and mothers. By 1960, television had displaced the radio from the living room with eighty-five percent of American households owning a television set (US Census Bureau, 1999).[3]

Television's takeover of the assemblage, particularly in the North American context, would not be complete until it had overcome technical limitations in relation to transmission. Unlike radio frequencies, those used by television require direct line of sight, thus necessitating the need for each home to have an aerial antenna for reception. Over-the-air (OTA) delivery systems work best in densely populated, geographically flat areas: in the United Kingdom, for example, eighty-one percent of households could receive OTA TV by 1951 (Branston & Stafford, 2010). Engineers knew that this limitation could be overcome through the use of coaxial cable, originally invented to improve telegraph transmission. The first transmission of an experimental broadcast via coaxial cable took place in Germany in 1936 for the Berlin Olympics (Early Television Museum, 2017). The first coaxial cable in the United States was laid between New York and Philadelphia in 1937 and within five years consisted of an 800 mile circuit (*Ibid.*). CATV or community antenna TV was first used in the United States and Canada in terrains where OTA reception was very poor. Large community antennas were set up and coaxial cable then connected individual homes to the antenna (Morrison, 2014). In 1952, there were seventy cable systems in the United States servicing 14,000 subscribers (CCTA, 2016). Ten years later, entrepreneurs and then large corporations saw the potential of the technology to capture and deliver a wide range of broadcast signals far beyond the local broadcast range, and began to offer multichannel packages for higher fees (CCTA, 2016).

The Analog Era

From the mid-1960s to the mid-1980s, television's programming and broadcast schedule consolidated into what is now understood as the classic network era (Lotz, 2009). Williams (1975) notes that while Western, crime, and medical dramas were already popular on radio and in print, television took these genres to new heights of popularity. American series, including the prime time serials such as *Dallas* and *Dynasty*, continued to dominate the global television trade, with the bulk going to seven countries: Australia, Canada, France, Germany, Italy, Japan, and the United Kingdom (Barker, 1999). By the 1990s, almost half of all imported programming in Western Europe came from the United States (*Ibid.*).[4] Outside of prime time, local affiliates and the independent stations bought syndicated programming to fill in "fringe" slots in the mornings and afternoons as well as on weekend afternoons. Derek Kompare describes syndication as "the primary mode of non-network distribution of television content," both first-run and rerun (2009, p. 57). Syndicated television is "almost nobody's favorite form"—"just-see" as opposed to "must-see" TV (p. 55)—yet it plays an equally important role as network television does in the creation of broadcast *flow*. Williams (1975) was the first scholar to recognize and characterize the programming day as such. What appears to be programming made up of discrete units marked by interruption, whether by commercial breaks, news updates, and/or public service announcements, has its own internal organization and coherence. This flow is what distinguishes the experience of watching television from watching a film in the cinema or reading a book. Fiske further develops the concept, describing the televisual flow as "discontinuous, interrupted, and segmented" (1987, p. 105). He also recognizes the unevenness of flow across broadcast models. The most discrete and coherent programming flow was found in the former Eastern bloc countries, followed by those of the United Kingdom, Australia, New Zealand, and the Western European countries, and most segmented found in the United States and Canada (*Ibid.*).

The television assemblage established its first linkages with participatory culture at this time as well. Fan clubs devoted to popular music and movie stars date back at least to the 1930s (Theberge, 2005). The first fan convention, devoted to science fiction, was held in 1936 (Bacon-Smith, 1992). Early television fandoms formed around *Star Trek: The Original Series* (US) and *Doctor Who* (UK). The official Star Trek Fan club was set up by the series creator Gene Roddenberry in 1968 and lasted one year until the series was cancelled. Fans had engaged in a successful letter-writing campaign to get the series renewed for a third season but there was to be no fourth. Through syndication, however, its popularity only increased, with the LA Times describing its ratings as "enviable" (cited in Pearson, 2011, p. 121). The first *Star Trek* fanzine was produced in 1967 (Coppa, 2006), and the first fan convention was held in New York in 1973 (Bacon-Smith, 1992). *Doctor Who*, which has remained in production since its debut in November 1963, also had an official fan club from this time period, which became the Doctor Who Appreciation Society in 1976. *Doctor Who* is also one of the British series that is sold internationally as part of the "reverse flow" of English-language programming imported to the United States (Barker, 1999).

Television's first reassemblage began in the 1970s in the United States when cable technology in combination with new satellite technology and deregulation enabled Charles Dolan and Gerald Levin to set up Home Box Office (HBO) out of New York, and Ted Turner to set up the superstation WTBS in Atlanta. Cable expansion continued rapidly into the next decade, resulting in what is often referred to as the "multichannel universe" with an increase from 16 million to 53 million household subscriptions (CCTA, 2016). By the early 1990s, over ninety-seven of US subscribers had more than thirty channels (Uricchio, 2004). Demand for more programming in turn led to the creation of new commercial networks: Fox in 1986, and the WB and UPN in 1995. Roberta Pearson (2011) contends that it was the success of *Star Trek: The Next Generation* that led to the launch of the latter. Many of the new channels became specialty content channels for niche audiences, beginning with twenty-four-hour news networks such as CNN, which served to reframe the programming flow as a continuous cycle (Fiske, 1987). By the early 1990s, direct satellite broadcast technology had the capacity to provide an alternative delivery system to cable. As a result, the expansion of the multichannel universe to countries where cable remained marginal or nonexistent was made possible.

TV sets became both larger and smaller and were added to other spaces in the home such as bedrooms, kitchens, and recreation rooms. It was the addition of two ancillary devices, however, that was responsible for "a shift away from the programming-based notion of flow that Williams documented to a viewer-centered notion" (Uricchio, 2004, p. 168). The first was the remote control device. The technology dates back to the early days of radio and was a novelty gadget in relation to television (e.g., Zenith's Flash-Matic and Space Command), only becoming ubiquitous in the 1980s with the advent of coded infrared devices (*Ibid.*). Enabling the viewer to "channel surf" or "zap" was not simply a matter of convenience but a necessity in managing the number of channels on offer (Thomas, 2011). The second ancillary device was the home videocassette recorder (VCR). The first consumer (albeit short-lived) Sony Betamax recorders were sold in Japan in 1975, followed a year later by the JVC VHS recorder (Gibbs, 2015). Sales of VCRs in the United States exceeded one million units by 1979 (Quigley, 1992). Ten years and 80 million units later, almost two-thirds of US households had a VCR (*Ibid.*). The percentage was even higher in the United Kingdom in households with children under sixteen at almost seventy percent (*Ibid.*). The practice of viewing recorded television programming at a time other than that of broadcast has come to be known as time-shifting, a term that dates back to the Sony Betamax US marketing campaign (Greenberg, 2008). As David Gauntlett and Annette Hill argue, users are able to "harness the technology for the ways in which it can be deployed to contribute to one's own life, rather than just using it in the ways proposed in the instruction manual" (1999, p. 146). The VCR thus allowed for programming to be slotted into one's routine rather than organizing one's routine around the broadcast schedule.

The home video market that the VCR provided for post-theatrical film release, however, could not be replicated for television due to technical limitations: the tapes only allowed for two hours of good-quality playback (four and six hours were possible for home recording but the quality was much lower). A single episode was therefore too short and an entire series too long to be commercially viable: one solution was a "Best of" set of classic episodes. According to Kompare (2006), the first such set was *The Honeymooners*. Memory Alpha, a Star Trek wiki, states that Paramount Home Video released ten episodes on a five-tape set in both VHS and Betamax format after the release of the first *Star Trek* movie in early 1980. Taken together, both home and commercial recordings served to decouple the television text from the broadcast flow, turning the ephemeral text into a discrete text and material object.

The VCR also enabled several fan practices. First it allowed for the creation of a personal archive of recordings which in turn allowed fans to rewatch as many or as few episodes at a time as often as desired. According to Jenkins, texts "accumulate" meaning through repeat use (1992, p. 51). Moreover, video tapes could be shared among fans: "the exchange of video tapes has become a central ritual of fandom, one of the practices helping to bind it together as a distinctive community" (p. 71). The VCR also enabled the practice of viewing multiple episodes in a row. Jenkins provides examples of communities hosting marathon viewings at a local host's home. As with syndication, video exchange could lead to discoveries of new fandoms: Jenkins mentions how he and his wife became fans of the British series *Blake 7* through the gift of the first season on tape. His personal example is also an early illustration of how fan culture creates a sharing or gift economy (Booth, 2010) outside national and commercial television markets. Finally home recording technology enabled the production of fan videos or vidding. According to Coppa (2006), the first fan video (vid) came out of *Star Trek* fandom and was made around 1980 using two VCR decks.[5]

The Assemblage Goes Digital

Digitalization at its most basic technical level is the translation of information into binary digits (zeros and ones). It is most commonly associated with the ICTs that enabled the formation of the internet and computer-mediated communication. Participatory fans were among the early internet adopters, creating cyberspaces in which to interact with other fans (Bury, 2005). SF [Science Fiction]-LOVERS was one of the earliest listservs dating back to the ARPANET of the 1970s (Sterling, 1993). GEnie, a rival to CompuServe that was set up in 1985, had "roundtables" dedicated to various fandoms (Busse & Hellekson, 2006). According to Nancy Baym (2000), rec.arts.tv.soaps was one of the oldest newsgroups on Usenet, a pre-internet service set up in 1980. The digitalization of television, however, did not begin until the mid-1990s. Unlike an analog signal, the digital signal suffers no degradation and is multiplexed, i.e., audio and video are processed together. Moreover, because a digital signal can be compressed without losing information, even more channels, both standard definition (SD) and high definition (HD), can be transmitted on the same frequencies (Morrison, 2014). The result was availability of even more channels via digital terrestrial television (DTT) delivery systems, not just via cable and satellite. The CBS affiliate WRAL became the first station to broadcast in HD in 1996 (WRAL, 1996). The first fully digital terrestrial service, ONdigital, was launched in the United Kingdom in 1998 (ONdigital, 2002). In the early 2000s, most developed countries began to prepare for a full digital switchover: The Netherlands became the first to complete the switchover, followed by Germany in 2008, United States in 2009, Australia, Canada, and New Zealand in 2011, and the United Kingdom in 2012. With the development of the HDTV set, which has adopted the display technologies of computer monitors, television is no longer the "insufficient medium of visual broadcasting" as it was characterized by Williams (1975, p. 28).

Digitalization has resulted in convergence, that is, the merging of discrete devices and systems into a singular device or system (Jenkins, 2006). Upgrades to networks using fiber optic cable, along with regulatory changes in the US context, allowed cable providers to become internet service providers, effectively displacing free dial-up services accessed through telephone lines. Their monopoly on broadband did not last: telecommunications corporations ("telcos") also upgraded their networks to provide dedicated subscriber line (DSL) internet service; today both types of providers offer what the industry calls a "triple play": internet, television, and telephone, "all on a *single wire* into the home" (CCTA, 2016, emphasis mine). The home itself, however, is better understood as a site of digital *divergence* (Thomas, 2011). Upon entering the home, the cable connects first to an ASL or DSL modem and then to the TV set, requiring a set-top box to convert the signal from digital to analog for older TV sets. Although HDTVs are able to receive a signal directly, most providers require a set-top receiver to access the signals that have been scrambled to prevent unauthorized distribution.

As for devices connected to the TV, two digital devices are required to replicate analog VCR functionality. The first is the digital versatile disc (DVD) player, the standards for which were established in 1995 (Frankel, 2007). With compression enabling more episodes of a series to fit onto a single disc, television content producers and their broadcasting networks finally became players in the digital video home market. By 1998, approximately 1.4 million US homes had a DVD player (Frankel, 2007). By 2016, seventy-seven percent of American households had at least one (Nielsen Company, 2016a). The creation of a box set for an entire television season/series began with Fox's release of the first season of *The X-Files* in 2000, thereby linking television for the first time to "the publishing model of media production and distribution" (Kompare, 2006, p. 338). Three years later, sales of TV box sets had grown substantially. *Family Guy: Volume One*, for example, sold 1.6 million copies, enough for Fox to order more seasons (Frankel, 2007). Kompare (2006) provides a list of over seventy box sets released on DVD between 2000 and 2003, made up primarily of American series as well as a few BBC series. DVD box sets, Kompare argues, "provide the content of television without the 'noise' and limitations of the institution of television" (p. 352). Like VHS tapes, they turn television texts into discrete objects removed from the broadcast flow. Unlike home recordings, DVD sets often include "valued added" extras such as deleted scenes and/or commentary by writers and actors. In short, "they present their series complete, uncut, organized, pristine, and compact, all qualities sought by … collectors" (p. 352). Like videotapes, DVDs can also be rented; in 1998 Reed Hastings set up Netflix as a mail-order DVD-rental business to compete against the video rental stores and chains such as Blockbuster.

The second ancillary device to replace the VCR is the digital video recorder (DVR), sometimes referred to as a personal video recorder (PVR). The first such service/device to launch was TiVo in the United States in 1999. Unlike the VCR, the internal hard drive allows the pausing and playback of live television, enabling another means by which the broadcast flow is altered. While unpopular with commercial television networks for obvious reasons (TiVo allows the deletion of ads in their entirety), DVRs were soon embraced by cable companies and their telco rivals, who began offering their own integrated devices for an additional fee or free as part of their higher-tier subscription packages. The percentage of American households with a DVR tripled between 2007 and 2011 (Nielsen Company, 2009) and reached fifty-three percent in 2016 (Nielsen Company, 2016b). Ofcom (2015) reports that sixty-four percent of UK households had one in 2015. Australia was not far behind at fifty-eight percent (Think TV, 2016). DVR penetration in Canadian households was 52.1 percent as of 2014 (Television Bureau of Canada, 2014).

The Rise of IPTV

The convergences and divergences discussed earlier were the springboard for a television-internet hybridization, which led to the development of IPTV. In the early 2000s, internet protocols became advanced enough to compress and encode video. Those users with broadband could upload and download files via peer-to-peer (P2P) file-sharing networks or Apple's iTunes service. Users could also stream these files from a server using software such as RealPlayer, Apple's Quick Time, and Windows Media Player. The first television content was illegally distributed online by fans using a BitTorrent client in 2003

(Schiesel, 2004). By piggybacking on the public internet infrastructure, such unauthorized downloading or "piracy" breaks completely with the foundational broadcast model based on centralized transmission. Playing catch up, the networks, with ABC (US) leading the way, made some content available for purchase on iTunes in 2005. ABC was also the first network to make full-length episodes available for streaming from its website to promote its new series in May–June 2006. It began streaming its regular schedule of dramas and comedies beginning with the 2006–2007 season. Hulu, a dedicated site for the streaming of television content, was launched as a joint venture between NBC Universal and News Corporation (Fox) in 2007. A similar service called TVCatchUp was launched in the United Kingdom and a year later the BBC introduced its iPlayer. Designed for secondary or complementary use to the original broadcast, only a few recent episodes of current series are generally available. The cable, telco, and satellite providers (henceforth referred to as cable+ providers as per the Nielsen Company) followed suit, offering their own video on demand (VOD) services that included television programming.

Until 2006, a computer was needed to download and stream television content. While P2P sharing still requires one, the release of AppleTV created the ability to stream content directly to a TV set. Other media streaming devices by Google, Roku, and Boxee soon followed, and streaming capabilities were added to gaming consoles such as Xbox360 and Nintendo Wii. In 2007, Netflix began offering a free "value-added" streaming service to its existing mail-order subscribers (Netflix, 2016). The company also began its expansion outside the United States in 2010 to Canada. As of 2016, Netflix claims to be "worldwide," by which it means it is available in 190 countries and has 74 million subscribers (Jackson, 2016). As such it is on its way to becoming "the first global TV network" in the words of its founder and CEO (cited in Jackson, 2016), and it effectively serves as a transnational model of broadcasting. It also became the first non-network producer and distributor of serialized drama in 2013 with *House of Cards* and now produces dozens of series in several genres, including drama, comedy, and children's programming.

In addition to television and computer screens, mobile devices such as smart phones and tablets can be used for online viewing. Cellular telephony has been around for a number of years but the capacity for such devices to function as a screen for the viewing of television programming only dates back to 2005, when Korea launched a digital multimedia broadcasting system (Ghadialy, 2006). Apple's first iPhone was released in 2007, which came with apps for iTunes and YouTube. In 2010, Netflix released apps for the iPhone, iPod, and the first iPad. Today all smart phones and tablets offer a range of apps for streaming services. According to Nielsen, eighty-one percent of Americans have mobile phones and fifty-eight percent have tablets (2016a). A survey by the Pew Research Center found that fifty-two percent of Americans had used their smart phone as a *second screen* while viewing television programming (Smith, 2012). The same percentage of Australians over the age of fourteen have a cell phone and forty-nine percent have a tablet (Oztam & Nielsen Company, 2016). Almost half of Canadians own a tablet, and two-thirds of Canadians own a smart phone (CRTC, 2015). Almost one-third of Canadian smart phone subscribers have used their phone while watching TV, and thirty-five percent who own a tablet have done the same. Four out of five adults in the United Kingdom have a smart phone and just under two-thirds had a tablet in 2015 (Lee & Talbot, 2016). The Pew Research Center reports a lower percentage of smart phone ownership than the studies listed above: sixty-eight percent for the United Kingdom and seventy-two percent for the United States (Poushter, 2016). Of the Western European countries, the percentage of smart phones was highest in Spain (seventy-one), with Germany at sixty (Poushter, 2016). Although portable televisions have existed since the 1970s, smart phones and tablets serve to untether television, quite literally, via cellular and wireless technology. "It is perhaps more appropriate," argues Dawson, "to think of television as *site unspecific*," enabling texts to move between platforms and environments (2007, p. 238).

Rhizomatic TV

Since the late 2000s, a steady stream of media reports have suggested that IPTV is the future of television, pointing to the increasing numbers of "cord cutters" and "cord-nevers" i.e., those cancelling cable subscriptions or not subscribing to cable at all (see Greenfield, 2012; Harris, 2015; Lewis, 2015). In addition, a decline in live viewing and the increasing popularity of online viewing is cited particularly among a younger demographic (see Lawler, 2010; Schonfeld, 2010). Data from a range of industry, government agency, and marketing research sources do indeed provide evidence of these shifts. Yet they also show just how entrenched the services and modes of viewing associated with BTV are. Regarding receipt of a broadcast signal, Nielsen (2016a) pegs US household cable+ penetration at eighty-five percent, with twelve percent only receiving a broadcast signal and three percent receiving broadband only.

In Canada, ninety-three percent of Canadian households have a cable+ subscription, eighty-five percent of which are digital (CMS, 2015). Germany has the highest number of cable customers in the EU (6+M), double that of the United Kingdom and the Netherlands (3+M) (Cable Europe, 2014). The Pew Research Center reports that fifteen percent of Americans are cord cutters and nine percent could be considered "cord nevers" (Horrigan & Duggan, 2015). About sixteen percent of Canadians do not pay for cable, half of whom can be considered cord cutters (CBC, 2015). A more recent study says that cable+ providers lost almost 100,000 subscribers between March and September 2016, thirteen percent more than during the same period in the previous year (Harris, 2016). Cord cutting is of course a practice that is only meaningful in national contexts where cable+ services are dominant. For example, despite having the second highest number of subscriptions in the European Union, cable penetration in the United Kingdom was about fifty-one percent in 2015 (Ofcom, 2015). It is much lower in Australia at twenty-nine percent where ninety-seven percent of all homes receive DTT (Oztam & Nielsen Company, 2016).

The amount of decline in live viewing reported varies depending on the source. According to Nielsen, American adults spent an average of four hours and thirty-one minutes a day watching live TV in 2016, only down a percentage point from 2014 (2016a). Time-shifting on the other hand accounted for just over half an hour of daily viewing. In contrast, Hub Entertainment Research found that only forty-seven percent of adult Americans watched live TV and the rest was time-shifted, a term used broadly in this particular study to include use of a DVR (thirty-four percent) and a Pay TV VOD service (nineteen percent) as well as subscription streaming services (Loechner, 2015). Ofcom (2015) reported that eighty-five percent of UK households viewed television live in 2014, a decline of only seven percent since 2010.

In relation to online viewing, Hub Entertainment reported that sixteen percent of adult Americans streamed television content from Netflix, eight percent from TV network sites or apps, six percent from Hulu or Hulu Plus, three percent from Amazon Prime or Amazon Instant, and seven percent from other online sources (Loechner, 2015). Media Technology Monitor found that almost half of adult Canadians living outside Quebec viewed programming online, with one in ten saying that they streamed and downloaded all their TV content. Just under forty percent were Netflix subscribers (Oliveira, 2015). In the United Kingdom, the BBC iPlayer dominates TV streaming across all devices at thirty-one percent, followed by Sky at sixteen percent (Ofcom, 2015). Ofcom also reported that 4.4 million UK households subscribed to Netflix. Netflix and other streaming services were only launched in Australia in early 2015 and the Australian Communication and Media Authority (2015) reported that six months later, seventeen percent of viewers were using them, but all three percent in relation to Netflix.

Since audience measurement companies do not track unauthorized downloading and streaming, we need to look elsewhere for statistical evidence. According to Michael Newman (2012), television accounts for half the BitTorrent downloads. The web publication TorrentFreak has reported that the most pirated show since 2011 is HBO's *Game of Thrones* (Van Der Sar, 2016). PricewaterhouseCoopers surveyed just over 200 participants who admitted to pirating copyrighted media content. Over eighty percent said that they had streamed TV shows for free in the past six months and sixty-one percent said that they had downloaded shows for free (Bothun & Lieberman, 2011).

An Ipsos OTX survey (2014) provides the most comprehensive snapshot of viewing modes not only because of its large sample of over 15,000 respondents but because of its international scope (twenty countries). It confirms the findings of the industry and national regulatory bodies that the large majority (eighty-six percent) of viewers engage in the traditional live mode. It also found that almost a third (twenty-seven percent) watched programming on a computer. Time-shifted viewing through use of a DVR or streaming via a set-top device connected to a TV set were tied at sixteen percent, and just over ten percent watched TV on a mobile device.

Rather than rely on a linear logic of progress to assume that IPTV is on a steady course to replace BTV, the processes of hybridization are better understood as *rhizomatic* (Deleuze & Guattari, 1987). Like a rhizome in nature, assemblages are neither "reducible neither to the One or the multiple ... but always a middle (*milieu*) from which it grows and overspills" (p. 23). BTV and IPTV are thus better understood as *intra-assemblages* (Ibid.), with signals, devices, and modes of viewing that are distinct yet overlap at the site of reception.

Concluding Thoughts

To sum up, television has a long history of assemblage and reassemblage that preceded its hybridization with the internet assemblage. Indeed much of what we associate with television preceded its invention: the national broadcast model of centralized transmission and private reception, the networks, the programming day, seriality, and audience measurement. When television arrived in the home, it was already domesticated, its set steadily replacing that of radio in the living room. Histories of television have tended to position radio as "before TV" but until the 1960s, it is more accurate to think of television as "after radio." Cable expansion, particularly in Canada and the United States, and the development of satellite technology played a key role in creating the analog multichannel universe, which in turn enabled the production of more series, for not only prime time but also syndication and export. The introduction of the remote control and VCR was responsible for the shift from network-centered to viewer-centered flow, no longer requiring household members to organize their routines around the broadcast schedule. Thus it is fair to say that both live and time-shifting modes form the basis of contemporary BTV reception.

Digitalization also led to a series of convergences and divergences outside and inside the home. While a single wire has the capacity to carry both a cable and internet signal, and both cable and telecom providers offer both services, the number of devices associated with television viewing includes DVRs, DVDs, set-top boxes, streaming devices, computers, and mobile devices. DVD technology in particular added an additional node to the assemblage, turning television into a commercial, material, collectable object, and one which served as a springboard for streaming and its associated practices. Moreover, while popular media fan practices and participatory culture exist separately from the television assemblage, numerous linkages have been established, beginning with local fan clubs and conventions associated with the *Star Trek* franchise and *Doctor Who*. The VCR enabled the creation and sharing of fan vids; the internet extended fan-related discussions and enabled the sharing of creative fan works in cyberspace.

Finally, the development of IPTV has resulted in the television assemblage becoming more rhizomatic; for the first time since the institutionalization of radio in the early 1920s, a broadcast signal is no longer required—transmission relies on the decentralized nodes of the internet instead. It also disrupts the national model of broadcasting, a model which served as the foundation of the first media age (Poster, 1995). The next chapter attempts to untangle and trace these rhizomatic, hybrid connections and look at user investments in both the BTV and IPTV intra-assemblages.

Notes

1. Advertisements from the time typically showed the console against the wall with a vase of flowers on top. For an example, see https://frrl.wordpress.com/2009/01/22/long-live-the-all-american-five-or-recovering-a-piece-of-radio-history/ and http://www.digitalhistory.uh.edu/disp_textbook.cfm?smtID=8&psid=2645&filepath=http://www.digitalhistory.uh.edu/primarysources_upload/images/listening_to_radio_l.jpg
2. The popularity of *Amos 'n' Andy* is both astonishing and disturbing given it was effectively a serial minstrel show, with white actors doing "blackvoice."
3. Radio also had a place in the suburbs. As part of its reassemblage, radio miniaturized and multiplied, ending up in kitchens and bedrooms, and then followed Elvis out the building so to speak, imbricated in new cultures of rock music and mobility. Digital convergence and cellular/wireless technologies only enable television to catch up fifty years later.
4. It is noteworthy that the first book-length empirical study on the reception of a popular television text was conducted by Ien Ang (1989) with Dutch fans of *Dallas*.
5. Vidding at this time was both expensive and labor intensive. Coppa provides an analysis of a well-known 1990 "meta" vid about the process of creating a vid by the California Crew, set, appropriately, to the Billy Joel song, Pressure (see http://mediacommons.futureofthebook.org/imr/2008/01/28/pressure-metavid-california-crew).

References

208 Radio Luxembourg. (2001, February 7). History of Radio Luxembourg and its English service. Retrieved from http://www.radioluxembourg.co.uk/?page_id=2
ACMA. (2015). Subscription video on demand in Australia 2015. Retrieved from http://www.acma.gov.au/theACMA/engage-blogs/engage-blogs/Research-snapshots/Subscription-video-on-demand
Ang, I. (1989). *Watching Dallas: Soap opera and the melodramatic imagination* (D. Couling, Trans.). New York: Routledge.
Bacon-Smith, C. (1992). *Enterprising women: Television fandom and the creation of popular myth*. Philadelphia, PA: University of Pennsylvania Press.
Barker, C. (1999). *Television, globalization and cultural identities*. Philadelphia, PA: Open University Press.
Baym, N. K. (2000). *Tune in, log on: Soaps, fandom, and online community*. Thousand Oaks, CA: Sage Publications.
Bergreen, L. (1980). *Look now, pay later: The rise of network broadcasting*. New York: Doubleday.
Boddy, W. (1990). *Fifties television: The industry and its critics*. Chicago: University of Illinois Press.
Booth, P. (2010). *Digital fandom: New media studies*. New York: Peter Lang.
Bothun, D., & Lieberman, M. (2011, October). Speed of life: Discovering behaviors and attitudes related to pirating content. PricewaterhouseCoopersLLP. Retrieved from: http://www.pwc.com/us/en/industry/entertainment-media/publications.html

Bouckley, H. (2016, October 3). From Marconi and the transistor radio to DAB: The history of radio in the UK. Retrieved from http://home.bt.com/tech-gadgets/from-marconi-and-the-transistor-radio-to-dab-the-history-of-radio-in-the-uk-11364015764901

Branston, G., & Stafford, R. (2010). *The media student's book* (5th ed.). London: Routledge.

Bury, R. (2005). *Cyberspaces of their own: Female fandoms online*. New York: Peter Lang Publishing.

Busse, K., & Hellekson, K. (2006). Introduction. In K. Hellekson & K. Busse (Eds.), *Fan fiction and fan communities in the age of the internet: New essays* (pp. 5–32). Jefferson, NC: McFarland & Co.

Cable Europe. (2014). Cable facts and figures. Retrieved from http://www.cable-europe.eu/industry-data/

Carlat, L. (1998). "A cleanser for the mind": Marketing radio receivers for the American home, 1922–1932. In R. Horowitz & A. Mohun (Eds.), *His and hers: Gender, consumption, and technology* (pp. 115–137). Charlottesville: University Press of Virginia.

CBC. (2015). Growing number of Canadians cutting traditional television CBC research shows. Retrieved from http://www.cbc.ca/news/business/growing-number-of-canadians-cutting-traditional-television-cbc-research-shows-1.3139754

CCTA. (2016). History of cable. Retrieved from https://www.calcable.org/learn/history-of-cable/

CMS. (2015). Canadian media statistics. Retrieved from http://canmediasales.com/canada-101/canadian-media-stats/

CRTC. (2015). Communications monitoring report 2015. Retrieved from http://www.crtc.gc.ca/eng/publications/reports/policymonitoring/2015/cmrre.htm#ex

Coppa, F. (2006). A brief history of media fandom. In K. Hellekson & K. Busse (Eds.), *Fan fiction and fan communities in the age of the Internet* (pp. 41–59). Jefferson, NC: McFarland & Co.

Dawson, M. (2007). Little players, big shows: Format, narration, and style on television's new smaller screens. *Convergence: The International Journal of Research into New Media Technologies, 13*(3), 231–250. doi: 10.1177/1354856507079175

Deleuze, G., & Guattari, F. (1987). *A thousand plateaus: Capitalism and schizophrenia* (B. Massumi, Trans.). Minneapolis: University of Minnesota Press.

Douglas, S. J. (1999). *Listening in: Radio and the American imagination, from Amos 'n' Andy and Edward R. Murrow to Wolfman Jack and Howard Stern* (1st ed.). New York: Times Books.

Early Television Museum. (2017). Mechanical television. Retrieved January 15, 2017 from http://www.earlytelevision.org/bell_labs.html

Fiske, J. (1987). *Television culture*. New York: Methuen & Co.

Frankel, D. (2007). DVD timeline: Looking back at the format's history. Retrieved from http://variety.com/2007/digital/features/dvd-timeline-1117963613/

Gauntlett, D., & Hill, A. (1999). *TV living: Television, culture and everyday life*. London; New York: Routledge.

Ghadialy, Z. (2006). Mobile TV technologies. Retrieved from http://www.3g4g.co.uk/Other/Tv/Presentations/mobile_tv_introduction.pdf

Gibbs, S. (2015, November 10). Betamax is dead, long live VHS. *The Guardian*. Retrieved from https://www.theguardian.com/technology/2015/nov/10/betamax-dead-long-live-vhs-sony-end-prodution

Greenberg, J. M. (2008). *From Betamax to Blockbuster: Video stores and the invention of movies on video*. Cambridge, MA: MIT Press.

Greenfield, R. (2012). Forget cord-cutters: Cable companies should worry about cord-nevers. *The Atlantic*. Retrieved from http://www.thewire.com/technology/2012/08/forget-cord-cutters-cable-companies-should-worry-about-cord-nevers/55380/

Gripsrud, J. (2004). Broadcast television: The chances of its survival in the digital age. In L. Spigel & J. Olsson (Eds.), *Television after TV: Essays on a medium in transition* (pp. 210–223). Durham, NC: Duke University Press.

Harris, S. (2015, November 9). Cord-nevers could be bigger threat to TV than cord-cutters. CBC. Retrieved from http://www.cbc.ca/news/business/cord-nevers-cord-cutters-tv-1.3308072

Harris, S. (2016, November 16). $25 basic TV can't stop customers from cutting their cable in record numbers. CBC. Retrieved from http://www.cbc.ca/news/business/basic-tv-cord-cutting-cable-1.3847342.

Hilmes, M. (1993). Invisible men: *Amos n Andy* and the roots of broadcast discourse. *Critical Studies in Mass Communication*, 10(4), 301–321.

Horrigan, J. B., & Duggan, M. (2015). Home broadband 2015. Pew Research Center. Retrieved from http://www.pewinternet.org/2015/12/22/2015/Home-Broadband-2015/

Ipsos. (2014). Changing TV habits. Retrieved from http://ipsos-na.com/news-polls/pressrelease.aspx?id=6433

Jackson, J. (2016, Thursday January 7). Netflix: From DVD rentals to the verge of world domination. *The Guardian*. Retrieved from https://www.theguardian.com/media/2016/jan/07/netflix-streaming-ces-global-tv-network

Jenkins, H. (1992). *Textual poachers: Television fans & participatory culture*. New York: Routledge.

Jenkins, H. (2006). *Convergence culture: Where old and new media collide*. New York: New York University Press.

Kompare, D. (2006). Publishing flow: DVD box sets and the reconception of television. *Television & New Media*, 7(4), 335–360. doi: 0.1177/1527476404270609

Kompare, D. (2009). The benefits of banality: Domestic syndication in the post-network era. In A. D. Lotz (Ed.), *Beyond prime time: Television programming in the post-network era* (pp. 55–74). New York: Routledge.

Lawler, R. (2010). One-third of US adults skip live TV. Retrieved from http://gigaom.com/video/one-third-of-us-adults-skip-live-tv-report/

Lee, P., & Talbot, E. (2016). There's no place like phone: Consumer usage patterns in the era of peak smartphone. Deloitte LLP. Retrieved from http://www.deloitte.co.uk/mobileUK/assets/pdf/Deloitte-Mobile-Consumer-2016-There-is-no-place-like-phone.pdf

Lewis, M. (2015, April 13). Canadians scrapping cable packages in larger numbers. *Toronto Star*. Retrieved from https://www.thestar.com/business/2015/04/13/canadians-scrapping-cable-packages-in-larger-numbers-report.html?referrer=http%3A%2F%2Ft.co%2FgGb kX0KJL4

Loechner, J. (2015). Time shifted TV viewing is the default. Center for Media Research. Retrieved from http://www.mediapost.com/publications/article/247581/time-shifted-tv-is-the-default.html

Lotz, A. (2009). What is U.S. television now? *The Annals of the American Academy of Political and Social Science, 625*(September), 49–59. doi: 10.1177/0002716209338366

Marling, W. H. (2006). *How "American" is globalization?* Baltimore, MD: The John Hopkins University Press.

Matelski, M. (1995). Resilient radio. In E. C. Pease & E. E. Dennis (Eds.), *Radio: The forgotten medium* (pp. 5–14). New Brunswick, NJ: Transaction Publishers.

Morrison, S. (2014). The evolution of television delivery. Retrieved from http://leightronix.com/blog/the-evolution-of-television-delivery/

Nachman, G. (1998). *Raised on radio*. Los Angeles: University of California Press.

Netflix Media Center. (2017). About Netflix. Retrieved from https://media.netflix.com/en/about-netflix

Newman, M. Z. (2012). Free TV: File-sharing and the value of television. *Television & New Media, 13*(6), 463–479. doi: 10.1177/1527476411421350

Nielsen Company. (2009, April). How DVRs are changing the television landscape. Retrieved from http://www.nielsen.com/us/en/insights/news/2009/how-dvrs-are-changing-the-television-landscape.html.

Nielsen Company. (2016a). The Nielsen Total Audience Report: Q1 2016. Retrieved from http://www.nielsen.com/us/en/insights/reports/2016/the-total-audience-report-q1–2016.html

Nielsen Company. (2016b). The Nielsen Total Audience Report: Q4 2016. Retrieved from http://www.nielsen.com/us/en/insights/reports/2017/the-nielsen-total-audience-report-q4–2016.html

Ofcom. (2015). The communications market 2015. Retrieved from https://www.ofcom.org.uk/research-and-data/multi-sector-research/cmr/cmr15

Oliveira, M. (2015, February 18). 1 in 10 English Canadians no longer watch TV, just web video: Poll. *Huffington Post (Canada)*. Retrieved from http://www.huffingtonpost.ca/2015/02/18/nearly-1-in-10-anglophone_n_6707736.html

ONdigital. (2002). ITV digital history. Retrieved from http://www.onhistory.co.uk/

Oztam, & Nielsen Company. (2016). Australian multi-screen report: Q 01 2016. Retrieved from http://www.thinktv.com.au/content_common/pg-reports.seo

Pearson, R. (2011). Cult television as digital television's cutting edge. In J. Bennett & N. Strange (Eds.), *Television as digital media* (pp. 105–131). Durham, NC: Duke University Press.

Poster, M. (1995). *The second media age*. Cambridge, MA: Polity Press.

Poushter, J. (2016). Smartphone ownership and internet usage. Pew Research Center. Retrieved from http://www.pewglobal.org/2016/02/22/smartphone-ownership-and-internet-usage-continues-to-climb-in-emerging-economies/

Press Association. (2005, October 11). A history of the license fee. *The Guardian*. Retrieved from https://www.theguardian.com/media/2005/oct/11/bbc.broadcasting1

Quigley, J. (1992). Videocassette recorders. In D. Ulloth (Ed.), *Communication technology: A survey* (pp. 161–166). New York: University Press of America.

Schiesel, S. (2004). File sharing's new face. *New York Times*. Retrieved from http://www.nytimes.com/2004/02/12/technology/file-sharing-s-new-face.html?src=pm

Schonfeld, E. (2010). Live TV is for old people: Time shifting and online make up nearly half of all viewing. Retrieved from http://techcrunch.com/2010/08/30/video-time-shifting-online-half/

Silverstone, R. (1994). *Television and everyday life*. New York: Routledge.

Smith, A. (2012). The 'rise' of the connected viewer. Pew Research Center. Retrieved from http://www.pewinternet.org/2012/07/17/the-rise-of-the-connected-viewer/

Spigel, L. (1992). Installing the television set: Popular discourses on television and domestic space, 1948–1955. In L. Spigel & D. Mann (Eds.), *Private screenings: Television and the female consumer* (pp. 3–40). Minneapolis: University of Minnesota Press.

Sterling, B. (1993). A short history of the internet. Retrieved from https://w2.eff.org/Net_culture/internet_sterling.history.txt

Sterling, C. H., & Kittross, J. M. (2002). *Stay tuned: A history of American broadcasting* (3rd ed.). New Jersey and London: Lawrence Erlbaum Associates.

Television Bureau of Canada (2014). Numeris media technology trends. Retrieved from http://www.tvb.ca/pages/BBM_MediaTechnologyTrends_htm

Theberge, P. (2005). Everyday fandom: Fan clubs, blogging, and the quotidian rhythms of the Internet. *Canadian Journal of Communication*, 30(4), 485–502.

Think TV. (2016). A quick look at Fall 2015. *The Quarterly*. Retrieved from http://www.thinktv.com.au/media/stats_&_graphs/library/tv_is_everywhere.pdf

Thomas, J. (2011). When digital was new: The advanced television technologies of the 1970s and the control of content. In J. Bennett & N. Strange (Eds.), *Television as digital media* (pp. 52–75). Durham, NC: Duke University Press.

Tomlinson, A. (1990). Home fixtures: Doing-it-yourself in a privatized world. In A. Tomlinson (Ed.), *Consumption, identity, and style: Marketing, meanings, and the packaging of pleasure* (pp. 40–51). London and New York: Routledge.

Uricchio, W. (2004). Television's next generation: Technology/interface/culture/flow. In L. Spigel & J. Olsson (Eds.), *Television after TV: Essays on a medium in transition* (pp. 163–182). Durham, NC: Duke University Press.

US Census Bureau. (1999). Selected communications media 1920–1998. (No. 1440). Retrieved from https://www.census.gov/en.html

Van Der Sar, E. (2016). Game of Thrones most pirated TV show of 2016. *TorrentFreak*. Retrieved from https://torrentfreak.com/game-of-thrones-most-torrented-tv-show-of-2016-161226/

Webster, J. G., & Phalen, P. F. (2009). *Mass audience: Rediscovering the dominant model*. New York and London: Routledge.

Williams, R. (1975). *Television: Technology and cultural form*. New York: Schocken Books.

WRAL. (2014, July 14). History of WRAL digital. Retrieved from http://www.wral.com/history-of-wral-digital/1069461/

· 2 ·

HOUSEHOLD ASSEMBLERS

Patterns of Multiscreen and Multimodal Viewing

This chapter examines Television 2.0's configuration in the contemporary household. The TV 2.0 survey data indicate almost universal engagement with both the BTV and IPTV intra-assemblages: eighty-seven percent received a broadcast signal (DTT or cable+) and all had internet access, with just over ninety percent having watched television on a computer. To map out the relationship between the two intra-assemblages, it is necessary to take into account the ways in which services, screens, and devices are assembled by household members as well as the modes of viewing in which they engage. To this end I will first present some descriptive and inferential statistics from the survey to look at the broader patterns of engagement and determine any demographic patterns based on gender, age, and region.[1] Next I turn to the interview data for a more fine-grained analysis. My analysis revealed a more uneven and complex engagement: in addition to those who primarily engage with BTV and IPTV, the large majority are hybrid TV (HTV) assemblers who have no strong allegiance to either.

Intra-assemblage Relations

The survey asked respondents about their engagement with the television, computer, and mobile screen. The mean percentage for television was sixty, with forty-five percent spending between eighty percent and hundred percent of their viewing time in front of a TV set. Half the respondents, however, reported also having connected computers or media streaming devices to their TV sets, twenty-two percent having done so frequently. A mean percentage of forty reported viewing television programming on a computer or laptop, with a spike of just over twenty percent of respondents at the ten percent mark. Moreover, a regression analysis shows that an increase in the amount of time spent viewing on a computer screen predicts a decrease in time spent viewing on a television screen.[2] The mean percentage for use of a mobile device for viewing was much lower, at just under four percent.[3] There are demographic differences of significance as well. The mean TV/computer ratio of sixty/forty drops to almost fifty/fifty for the eighteen to twenty-nine cohort and to fifty-three/forty-eight among those residing in continental Europe. In contrast, the ratio increases to approximately seventy-five/twenty-five for those fifty and older. The UK respondents had the highest TV/computer ratio at sixty-five/thirty-five. Gender was only statistically significant in relation to connecting devices to a TV set, with more men doing so than women (mean score = 1.38 compared to 0.91 on a 4-point scale, wherein 2 = *sometimes* and 1 = *occasionally*). Taken together, these findings suggest that television remains the primary screen but the computer has come to rival it among Europeans and younger viewers.

As for modes of viewing, the mean for live viewing was thirty-two percent (far lower than the industry findings reported in Chapter 1). Over half the respondents reported that they watched twenty percent or less of their programs at the time of broadcast. The least amount of live viewing took place amongst the younger cohorts, with the biggest gap between eighteen and thirty-nine year olds (twenty percent) and those fifty and over (forty-seven percent). This finding is consonant with the industry studies. As for time-shifted viewing, the mean percentage was thirty-nine. Although age was not statistically significant for this mode, it was for DVR use: twenty-nine percent of those under thirty recorded programming using a DVR compared to a range of forty-five percent to sixty percent for those over forty; thirty-six percent was the mean percentage. The United States and Australia/New Zealand had the highest percentages of time-shifting (forty-seven percent), yet DVR use for the latter countries only averaged one-third, likely meaning that the question about time-shifting was understood to include online viewing. The mean percentage for the UK residents was forty-four. The Canadian and European residents, in contrast, time-shifted less than one-third of their programming (29.9 percent and 27.9 percent, respectively), which corresponded to their lower DVR use (twenty-five percent and thirty percent). Regression analysis demonstrates that an increase in time-shifting strongly predicts a decrease in live viewing.[4] Respondents were also asked how frequently they watched commercially produced DVDs: seventeen percent said never and thirty-three percent said occasionally.

As for online viewing, TV 2.0 asked questions about the frequency of streaming from network sites, as well as streaming from third-party services such as Netflix or YouTube.[5] The frequencies for both types of streaming sites were almost identical; a slightly higher percentage of respondents had never used a third-party site (eighteen percent) as opposed to a network site (fourteen percent). One-third were frequent "streamers" of both types of services. Thirty-seven percent had VOD services through their cable+ subscription. Engagement with downloading via iTunes or BitTorrent clients was unevenly distributed: while twenty-seven percent of the respondents were frequent downloaders, thirty-seven percent reported that they had never downloaded content. There were no statistically significant gender differences in relation to streaming but men were more frequent downloaders. Just as the youngest cohort did the least amount of live viewing and the oldest the most, the reverse pattern emerged for online viewing. On the 4-point frequency scale, the means were as follows for those aged eighteen to twenty-nine: 2.10 (network streaming), 1.95 (third-party streaming) and 1.57 (downloading). In contrast, the means for those aged fifty and older were 1.36, 1.31 and .67 respectively.

There were also significant differences across regions. The United States dominated streaming from third-party sites as well as VOD. Similarly, streaming from network sites was more popular in the United Kingdom and Europe, with a mean point score of 2 on the 4-point scale. The United States was just slightly behind with 1.77. Canada and Australia/New Zealand's mean scores were significantly lower at 1.48. Canadian use of VOD, however, at thirty-eight percent was much higher than European use (twenty-one percent). As for downloading, Europe dominated with a mean score of 1.83, followed by Australia/New Zealand at 1.67. Canadian respondents did the least amount of downloading at 1.13, closely followed by the US-based respondents with a score of 1.14. Regression analysis predicts a statistically significant decrease in live viewing in relation to downloading, a marginally significant decrease in relation to third-party streaming but interestingly no decrease in relation to streaming from network sites.[6] Time-shifting is a stronger predictor of a decrease in live viewing than is downloading.

Taken together, these findings demonstrate that viewing associated with Television 2.0 is both multiscreen and multimodal but that there are two distinct patterns of engagement. Those who are primarily invested in BTV are in the older age cohorts and live in North America or the United Kingdom. They watch a combination of live and time-shifted programming and are likely to own a DVR to this end. They may also have and use VOD services. They do not download content but they may do occasional streaming from network and third-party sites that do not require a paid subscription (e.g., Hulu). Those with the strongest investments in IPTV are generally part of a younger demographic and are more likely to reside in Europe, or Australia/New Zealand. In addition to not having a broadcast signal, and therefore not watching live television, they are likely to do a combination of downloading and streaming on a regular to frequent basis, either on a computer and/or using a device connected to a TV set.

My analysis of the interview data, however, revealed that the boundary between the two categories is in fact *leaky* (Haraway, 1988). I classified fifteen participants as BTV "loyalists": all but one was female; resided in Canada, the United Kingdom, or the United States; and only one was under thirty. Only six participants were unequivocally IPTV "trendsetters" by virtue of receiving no signal. After close analysis I included nine additional participants in this category. They still had DTT or cable+ service in their household but did almost all of their personal viewing online. Four of these fifteen participants were male, five were under thirty (ten under forty), and only four resided in North America. Just over two-thirds, however ($n = 41$), did not fit into either category. I have labeled them HTV "assemblers." Rather than form a third coherent category of consistent and common practice, I contend that the HTV assemblers extend the ranges of the BTV and IPTV categories. In this regard, I draw on Don Ihde (1990), who conceptualizes the cyborg as a liminal case cutting horizontally across the four categories of human/technology relations rather than as an emerging fifth category.

BTV Engagement

Home Delivery

> We are a two-person, five-television household. We like television.
>
> —Mary

All the BTV loyalists need to be considered "cable+ keepers." Mary, quoted above, lived in a rural area in the United States and needed a satellite service to get reception. She had a subscription with DISH TV that offered around 300 channels. She and her husband watched "ninety-eight percent to ninety-nine percent" of television programming on one of their TV sets. Brooke and her family lived in a rural area in Canada and in the past they could only receive the national public broadcaster (CBC) over the air. At the time of the interview they were Bell satellite subscribers and enjoyed both the convenience and variety of programming available. Aimee, also a DISH TV subscriber, enjoyed the specialty channels that were not available via DTT, including HGTV and Starz. Similarly Margaret had an expanded basic package from Charter (US) as well as a sports package. As a self-described "news junkie" Daisy subscribed to an additional news package. Anne E., who resided in the United Kingdom, originally upgraded from DTT to Sky Atlantic satellite service because the package included a number of American channels such as FX as well as a DVR: "Sometimes I feel there is just too much television and my life is getting in the way of my television watching … since getting satellite television there is a lot available to watch." For those residing outside of Canada and the United States, premium cable+ service provided access to current American and British series that were not available on their national networks:

> We have cable television with a premium package because the basic package doesn't include HBO. We have the premium package not for HBO but yes, for the other, for 3 MovieCities channels about mostly movies and documentaries. (Lauchita, HTV)

> I just switched to Sky recently. So I have the regular packages. There is a separate package for English entertainment, for sport, for royal news, and for high definition channels. And there is a new set of channels which, for example, that Discovery launched or the other channels which they have now. (Sashin, HTV)

Similarly, Vera resided in the Netherlands and had a mid-level package from Viggo. She liked the fact that she was able to receive local Dutch channels as well as Fox, BBC, and some French, Flemish, and German programming. Taken together, these samples demonstrate the persistence of cable+ services because of the easy and convenient access to a wide range of local, specialty, and/or international content.

Some of the HTV assemblers were more ambivalent about their delivery system, however. Max, for example, could be considered a cable "downgrader." He used to have a top tier package but had reduced his package because he was doing more online viewing. That said, he was not prepared to get rid of cable altogether because he liked watching news and sports live as well as recording programming using his TiVo. Idoru only had cable because it was a good deal when bundled with their internet service: "That's why I subscribed actually for the TV just to get a good internet." Similarly, Jayne and her husband got a "ridiculous deal" on basic cable when bundled with internet. These cases recall the discussion of industry convergence in Chapter 1 wherein the different types of providers offer the same service bundles. In Buffy's case, the apartment she moved into came with Bell satellite service (Canada), which included a DVR:

> I actually share the cost with the upstairs neighbours. It's something I kept mostly because my schedule doesn't permit me to always watch TV when it's on. (Buffy)

Only one BTV loyalist expressed some dissatisfaction with her cable service from Virgin (UK):

> We ... constantly say we are going to switch to Sky but we never do because the cable is convenient and it's already sunk into the pavement and drilled through the front wall of the house. Okay, right we will carry on. (Virginia)

This quotation highlights the ties between the household and neighborhood infrastructure, which makes it more difficult to change to a different type of service.

Seven of the eleven UK participants used the Freeview service to receive sixty plus DTT channels. sixty plus DTT channels. With no monthly cost, it is not surprising that, with one exception, none of these HTV assemblers mentioned a desire to get rid of their service even though they primarily watched programming online. Tarsus, who did express ambivalence, noted that she only kept her box to watch news, sports, and special events. Penguin was an American who cut the cord in 2005 but set up a backyard antenna to ensure he had good OTA reception:

> I hadn't had an outdoor antenna since the 60s at least at this household, I was adjusting it with an associate out on the backyard and I said, 'Hey! The picture's great.' He said, 'Come out and look at this.' The antenna was flat on the ground. Because of my location, pretty much comparatively close to the main broadcast transmission power in the area and so I have to struggle *not* to get a good signal. (Penguin)

These samples serve as a reminder that having DTT does not necessarily signal an investment in the BTV intra-assemblage just as cutting the cord does not necessarily signal a divestment.

Is It Live or Is It Time-Shifted?

A total of six participants, mostly BTV loyalists but a few of the HTV assemblers as well, watched most of their programming live. Kim (BTV) did so out of habit from the 1990s when she would rush home to catch the NBC Thursday prime-time line-up, which included *Friends* and *ER*. Others were home on weekday evenings after work and saw no reason not to view live:

> My life schedule is pretty much open so I don't really have a problem catching things when they first run and if I do happen to miss something that I really wanted to see I could pick up on iTunes or watch on the website or eventually it will be on Netflix. (Bunny, HTV)

Others did not want to miss a new episode of a favorite series:

> I prefer to watch [*True Blood* and *Falling Skies*] live but I do TiVo them just in case something happens where I am away from home and can't get back in time. So the only time I ever watch them time-shifted is when I, for some reason, can't get home to watch them live. (LWR, BTV)

> I would much rather watch *Doctor Who* live and on the TV than wait the next day and watch it on the computer. (Zee, HTV)

Like LWR, Daisy recorded her favorite shows as a back-up in case she was interrupted, and Elly only used her DVR in the case of scheduling conflicts when two or more programs were broadcast at the same time on different channels. The relationship between live viewing and fannish viewing will be explored in more detail in Chapter 4.

The participants who preferred to time-shift programming almost all used DVRs to this end. Willow (HTV) was one of two participants who still used a VCR: "I use things until they break." Several participants talked about increased amounts of time-shifting as a result of getting a DVR:

> We are in a really fortunate position when we moved into this place. Our TV is included with our rent. So our landlord recently upgraded and asked us if we wanted a PVR. We had to pay a bit more of course, but we were like sure! So we just got that in the spring. I would say it's kind of rocked our world! ... That's so shocking to me because I have always been really—like certain shows, I really always felt like I had to watch them when they came on. Although like I said, I think now I still will watch it the same day it's on but just not at that exact time. (Tasha, BTV)

Similarly, Notesofwhimsey (BTV) noted that since getting a DVR with her Telus (Canada) subscription, she only had live TV on "for background noise." She also no longer did any catch-up streaming from websites, using the DVR to record episodes instead. This particular change of viewing pattern challenges the notion of linear progression associated with technological development and use, in that the participant moved from online back to time-shifting.

The next set of comments capture the shift from a broadcast flow to viewer-centered flow that is afforded by time-shifting:

> It used to be that if I was watching a show live I would tend to watch them in chunks on the same channel. So I would watch an 8:00 show on the same channel and now it's much more à la carte. We can mix and match from any channel at any time and create the order in which we watch shows. So there is not that sense of 'we will just let the TV feed it to us' but we are now picking what we want. (Camden, HTV)

> I will want to watch *The Vampire Diaries* and then afterwards I want to watch *Nikita* but then I have to think, 'No, I actually want to watch *Grey's Anatomy*.' I will watch that one live because I prefer to watch *Grey's Anatomy* more than I would want to watch *Nikita*. So I watch *Nikita* later. (Julianna, HTV)

Penguin (HTV) hooked up his own DVR to manually record his content. He talked about how time-shifting made his viewing more "efficient":

> Let's say on a Thursday night there might be 4 half-hour comedies (US). I could watch 4 comedies in the time it used to take to watch 3. So I am actually spending less *time* watching it but I am watching more television. (Penguin)

The decision made by some of the HTV assemblers to time-shift almost all programming needs to be understood as a practice of resistance:

> I watch TV on my time now. It's on my schedule, not on the network's schedule. Again, with the exception of sports I don't care when anything is on anymore. I only care when I have the opportunity to watch it. So if I sit down and I have 45 minutes then I watch a 30 minute program. If I have an hour then I watch a 60 minute program. (Douglas)

> I kind of like to watch what I want to watch whenever I have the time because some nights I am at Uni until 10:00 at night and other nights I will be home at 4:00. I don't like to schedule my life around the TV schedule. Although I used to. My friends and I always used to leave a lecture early to be home in time to watch *Lost*. (Julianna)

A strong dislike of commercials also led a number of BTV loyalists and HTV assemblers, including Camden, Tasha, and Bunny, to time-shift. The next comment points to a rejection of the text structure sanctioned by the producers:

> I have been taping *So You Think You Can Dance*. Actually, I find the PVR so good for that show because honestly it's so bad. I really only want to see the dancing. I don't care about all the other comments and the back story on the dancers. (Tasha)

As noted in Chapter 1, DVR technology allows for a "live pause":

> I will time shift by 20 minutes. Like last night was the series finale of *Law and Order Criminal Intent* which is the only *Law and Order* [series] that I watch. And we started watching it 20 minutes into it so we could fast forward commercials. (Joan, HTV)

As a parent, Tabatha (BTV) appreciated this feature: "We are doing a lot more of pausing live, more than watching catch-up." Similarly, Anne E. noted, "When I go to hotel rooms and I can't pause live television, it's *extremely* disconcerting!" Viewing that is "not exactly live" (Farah, HTV) or is only "slightly time-shifted" (Tabatha) thus blurs the boundary between the two modes, allowing one to have one's live viewing cake and to eat it too.

A Steady Stream

While the BTV loyalists did no downloading, most did a limited amount of streaming. Since no one in this category hooked up a computer or device to their TV set, those without VOD services provided by the cable+ provider viewed streamed content on a computer.

> I don't stream shows unless I happen to be in a hotel room that has a fast enough connection. So I normally don't do that. To watch shows on my computer I watch DVDs. (Mary)

Ellen mentioned streaming news on her computer when she was traveling. Tabatha and Elly found it too awkward and inconvenient given their home setups:

> The US shows that I watch I watch it on live TV just simply because I guess it's easier. I use a laptop computer so if I were going to watch stuff say at Hulu.com or some of the other sites that air TV shows, it's a matter of dragging out the laptop, setting it up, plugging it into the wall outlet, hooking up my DSL cable line, and all that. So it's kind of hassle. It is easier to watch on live TV. (Elly)

Some of the BTV loyalists with VOD used the service instead of a DVR as backup or catch-up:

> So I tend to, if I am not a watching television live or if I have missed a particular show that I wanted to watch, I will access the Comcast cable system video on demand and simply pull up the show and watch it at a time that is more convenient for me […]. I always make an effort if I know there's a serial program that I like to watch, to catch up on the episode that I've missed prior to the subsequent week's episode airing. (Kim)

Both Anne E. and Virginia used the BBC iPlayer through their cable+ services, although the latter was not happy with the interface provided through her cable provider. Nonetheless, she still preferred it to using a computer:

> We do watch it on the TV but we tear our hair out trying to find the programs because the whole menu system is completely appalling. We did used to do some catch-up on the computer but it is a smaller screen basically. (Virginia)

A few BTV loyalists engaged in streaming for the same reasons that others used a DVR or VOD:

> There are three [series] that I like: *The Vampire Diaries*, *Big Bang Theory*, and *Community*. All at 8:00. And for a while there were four with *Bones* on at 8:00. Some of them you could get on demand and I could record two of them but that still left one that I couldn't do. So for cases like that then I stream it … through the laptop. In case I forget or the programming failed. (Rene)

> Usually if I am going to watch television on a computer it is because I either was not aware that the show was going to be on and I missed it and then it's not available on any of the other time zones. (Brooke)

> As I said, I prefer the TV. Normally when I am doing it on the laptop it's a matter of because I have to stream it online. I will also use as another [form of] time shifting. If it's available on Demand with Charter, I will use that a lot as a first resort if it's still there. (Margaret)

To sum up, it is clear that BTV still appeals to a number of viewers because it provides easy access to a range of selected content via DTT or cable+ delivery. In addition, there is the convenience of being able to turn on the TV set to either view live or to organize and reorder the broadcast schedule to fit with one's daily routine. Streaming serves as an additional means through which to manage the broadcast schedule and resolve conflicts either in conjunction with a DVR and/or VOD or as a backup of another backup. Thus, the time-shifting and online modes are intertwined in ways not captured by the BTV/IPTV distinction.

IPTV Engagement

The lack of a broadcast signal is one of the defining aspects of the IPTV trendsetter. Peter P. was a Brazilian who moved to the United States. He described cable as "useless" given that the programming that he was interesting in viewing was available on various streaming services.

> I have Netflix. I can watch movies, documentaries, television shows and I like the old stuff so they have the old stuff I like. I also watch anime. ... You have Hulu. You have YouTube. You have Vimeo. You have ABC, Discovery Channel. So come on! The stuff that I like to watch are *free* in streaming and I don't mind watching online advertising. So why pay more? (Peter P.)

In Brazil he used to have to download illegally to get the same kind of content. Philippe (Canada) had only been using live television for background viewing before he canceled his cable:

> I used to watch television really as a third or a second but mainly as a third thing that was open in the house/the room and it was like there, open, and I would watch it sometimes and then go back to read or study or do other stuff or cook or whatever. It was like a member of the family that was there but I was not sitting and really watching it. (Philippe)

Heresluck resided in an area with very poor DTT reception so she was faced with paying for cable or not having a signal:

> Cable is really expensive and you are paying for a lot of stuff that in my case I wouldn't actually watch and that just seems silly to me. I buy DVDs of shows that I want to watch. I sort of object to paying for something twice like that especially since again, there is whatever, a million cable channels that I just don't care about. (Heresluck)

Her next comment gestures to the lack of value associated with much of the programming produced for a multichannel universe.

> Not having cable I can't just sort of turn on the TV and then 6 hours later find that it's 3:00 in the morning and I'm watching the Home and Garden network which I would do, frankly, if I had cable. So for me, it's better to choose to acquire TV rather than sort of having it constantly. (Heresluck)

Heresluck was the only one of the six participants without a signal who continued to do most of her viewing on a TV set:

> I don't really watch TV on my desktop machine; I use it to download TV and then I typically copy the episodes to a USB stick and my DVD player has a USB input. (Heresluck)

The other five did not have a television set at the time of the interviews. Peter P. viewed streamed and downloaded content on a combination of computer, iPad, and mobile phone. Anna B., a UK resident, not only "cut the antenna" but got rid of the TV set to avoid paying the license fee.

> We didn't have very good reception and there were lots of channels that we only got really poor reception and we were finding that there were fewer and fewer shows that we actually would watch. We really felt that the quality of the kinds of things that we wanted to watch was going down in this country. (Anna B.)

Similarly, George F. was an Australian who once lived in a household that had cable when the service was first being rolled out in the 1990s. She has since cut both the cord and antenna.

> We've got a massive monitor that we've plugged into our Xbox and our Play Station so they feed off the house network and they get the internet [...]. It's permanently set up in our lounge space. (George F.)

As for the remaining nine IPTV trendsetters who had a broadcast signal, almost all their viewing was done online either on a computer or with a computer or device hooked up to their TV. Annika had cable in her household but said that she only watched TV on rare occasions for events such as the Olympic Games. She did almost all her viewing on her computer. She primarily used her TV set in conjunction with a DVD player. Neither M nor GSM watched broadcast television anymore either but had cable services because of other members in their respective households: M lived with her brother and sister-in-law in Brazil and watched over eighty percent of her programming on her laptop in her room. GSM lived with his wife and their two sons in Israel. His sons were the ones who did most of the live TV viewing. Madeleine lived in a student dormitory in Germany where cable was included. However, because so few of the American series that she liked were on the broadcast schedule, she did most of her viewing on her computer. She had a TV set but did not admit to it because then she would have been required to pay a license fee.

The remaining five IPTV trendsetters had DTT, but again, it was rarely watched.

> I've got my old laptop set up as a home theatre PC. I am running XBMC [now Kodi] and that is just connected to the TV. Anything that we watch that is time shifted comes off of that and we watch it on the TV now.[7] (Louise)

Freda, a Canadian living in Australia, mainly listened to the sound of live TV coming from the living room while she was in the kitchen cooking but watched those series of which she was a fan on her laptop. The other three were Americans who had cut the cord but still wanted access to network television without the expense. As Sophie said, "We would rather spend the money getting really nice internet than buying cable." That said, she still preferred to stream from network websites and Hulu, making a similar point to that of Heresluck about falling into the habit of watching programming of little value or interest to her. For her part Knitmeapony avoided watching broadcast TV as a practice of resistance "to remove commercials and advertising from my life […]. When I really, really want to see something I can finagle my rabbit ears into doing what it does." Jake was an American who had become accustomed to having DTT while living with his family in New Zealand. When they returned to the US, they started with a basic cable package and then found they were not watching enough TV to warrant the expense, replacing it with an antenna. Nonetheless, he streamed most of his content on his laptop: "I usually just lay on the couch with it, pretty much."

Libby and Gene were HTV assemblers who also chose DTT over cable+ services. Both lived in the same large American city and were able to receive almost forty local network affiliates through a digital antenna used with software from a company called Eye-TV. The software processes the signals on a Mac mini for output to a television set or computer monitor. The software also enables both the outputting of recorded and streamed content to both the television and other computers in the house. As Gene put it, "it's all a mash up. Is it a computer, it is a television? I don't really know." Libby's comment underscores the importance of approximating a BTV viewing experience while using an IPTV intra-assemblage:

> I made a deal with my girlfriend that if I could get access to any television show that she wanted to watch then we could get a cleaning lady instead of cable […]. She is not quite as technically inclined as I am so I needed to make it more like watching TV and less like finding stuff on your computer. (Libby)

Although this setup made it easy to watch network television live, she described doing so as "strange behavior for me." Gene was not technically a cord cutter because a cable subscription was included in his condo fees. As he explained, his use of Eye-TV was an overt rejection of the commercial broadcast model and part of a personal politics of resistance:

> We don't ever sit down and watch a program because of broadcast schedules [...]. It's a pattern that extends to every aspect of our life. So if someone calls on the phone we say, 'Put us on your no-call list.' These marketers! And we do that. And if something comes in the mail that's marketing or advertising we, depending on how aggressive and annoying it is we either notify them through their website to remove us from their list or we stuff everything back into their return mailer and say take us off your list. So we basically, because politically we are environmentalists, we are progressive activists, and we see the invasion of commercial license to every aspect of daily living as not a way we want to live our lives. So our television decisions are part of the broader political stance we take. (Gene)

Several other HTV assemblers connected devices and computers to the main household TV set. Helen described using a long cord connecting the desktop to the TV, which also served as a monitor for the computer. PC users such as Khal used Xbox and Microsoft software to view streamed content on the TV. Revan and his wife had a TV in the living room which was hooked up to both cable and a PC. They also had TVs in their offices, one hooked up to a PC and other to an Xbox. Camden built his own PVR out of a PC to both view and record programming from broadcast television and stream Netflix on multiple devices:

> We have it on the computer that runs our PVR and we have it on a gaming console, we have Netflix. And I also have a device, it's Google TV. I don't know if it's a device that you guys have come across yet but it runs Netflix as well. So we can basically get it on all of our televisions and on our mobile phones we can get Netflix. (Camden)

Several participants including Bunny and Will used a Wii console. Corsac connected an external hard drive to his TV. Lauchita's comment below offers one of the main reasons for outputting content to view on the television set rather than a smaller computer or mobile screen:

> I have a computer plugged into a plasma TV. So the source is the computer not the TV but I watch it on the TV because it's better and it's a very big screen. I have a computer that is mostly devoted to TV. So the source where we get the programs is the computer but we use the plasma TV as the monitor of the computer because we prefer to watch it in HD and in a better wider screen and a better screen than the computer. (Lauchita)

The size of screen was the main reason given for not viewing online content on a mobile phone or iPod:

> It was okay for me to make the small shift from having a television at home, from watching a couple of hours to watching so many hours on my laptop and then on my desktop. But to go back to watching a lot of television on such a tiny, tiny, little smart phone—no, that's never going to work. (Julianna)

A preference for a larger screen is also related to social viewing and will be discussed in the next chapter.

Download This!

One distinction between IPTV trendsetters and the HTV assemblers quoted above is their engagement in the practice of unauthorized downloading.[8] Those residing outside of the United States and Canada, such as Anna B., George F., Freda, M, and GSM, did so because they were unable to access a favorite series in a timely manner through a national network, the original broadcast network website, or a streaming subscription service.

> We search for the current shows that we are watching, and download them and watch them ... *Weeds* has just started again. We always watch *Weeds*. (Anna B.)

> When I lived in Canada I did stream [*Pushing Daisies*] off of the CTV.ca website. I would watch it on CTV.ca on my lunch break while I was working as a secretary. If the legal streaming option is available I will use it. It's just that 95% of the time the show that I want to watch is not available in a legal streaming option, up-to-date in Australia. If I lived in the US I think I would be a pretty dedicated Hulu user. (Freda)

Both Freda and Lauchita (Argentina) also mentioned using a proxy server to access Hulu in addition to downloading. The Americans and Canadians who downloaded frequently were almost exclusively cord cutters with no signal who also wanted to view episodes in a timely manner.

> I will take my little 8 gig USB stick with whatever episode of *Castle* or not so much Friends this year because I have been watching at a friend's house, but *Castle* or *Doctor Who* or *Friday Night Lights* or whatever and take it out to the living room and actually watch on my TV through the DVD player with the USB input. (Heresluck)

Downloading is thus imbricated with affective relations and fannish viewing; this association will be discussed in more detail in Chapter 4.

A number of the IPTV and HTV assemblers were also frequent streamers. The majority resided in the United States, not surprising given their easy access to not only the American network sites but subscription streaming services.

> If a show is available through Hulu I would usually prefer to watch it through Hulu because they seem to do a better job than most of the networks sites [...]. If it's only available through the network site, I will watch it at the network site. (Sophie)

> We have a Netflix subscription. ... So that's been actually a big thing. For the kids, the children's programming all comes through that. They can get their shows and they are only 2 weeks later than they were broadcast live and so they are fine watching things that way. It's a chance for us to kind of either catch up on stuff that we may have fallen out of watching or try out stuff that may be a year or two old but we never had watched before but the entire catalogue is on Netflix. So we will watch a lot that way. (Camden)

Courtney (IPTV) streamed series such as *CSI* and *NCIS* from the network website but just as she refused to pay for cable, she would not pay for subscription services like Netflix: "I go to SideReel.com and watch it because it's free streaming with no commercials. If it's not there, Hulu is my second choice. Those are really the only two. I don't pay." Courtney's position can be understood as one of overt rejection of commercialized transmission whether by cable+ providers or by streaming subscription. Knitmethepony also struggled with the costs associated with the latter but came to a different conclusion:

> I used to be so anti Paywall. And then I got Netflix and then Netflix turned to streaming and suddenly like it was like a flipping revelation. I have always been very content to say, okay what's out there is out there and I can only get the last 3 episodes on Hulu so if I *really* need my fix, I will watch those last 3 episodes. It's one of those things that you don't know you need and then you have it *and* you *need* it. Absolutely! There are some things that don't stream on Netflix but they are on Hulu Plus. So then it's like, oh is it worth $8 a month? *Yes* it is completely worth it!! I have to have these things! (Knitmeapony)

In summary, levels of investment differ among and within IPTV trendsetter and HTV assembler groups. Those with the strongest investments in IPTV have no broadcast signal, no TV set, and engage in more downloading than streaming. Some HTV assemblers will stream regularly via paid subscription services but also will have a broadcast signal and do some live viewing, even if most likely as background, as well as time-shift.

The Future Is Hybrid

As I hope to have demonstrated, both the BTV and IPTV intra-assemblages are now firmly embedded in most contemporary middle-class households around the world. While almost all viewing involves multiple screens and all three modes of viewing, a number of users can be classified as BTV loyalists or IPTV trendsetters. The former tend to be older and reside in Canada, the United States, or the United Kingdom, whereas the latter are more likely to be under the age of thirty and reside in Europe, Australia, or New Zealand. The majority, however, are better understood as HTV assemblers. HTV is not a coherent category but a loose collection of multiscreen users and multimodal viewers, some of whom are more aligned with the BTV intra-assemblage on account of their limited use of streaming for catch-up, backup, and scheduling conflicts, both broadcast and personal. Similarly, those who are more closely aligned with IPTV do not do much live viewing but will both time-shift and stream content regularly, use laptops, and connect streaming devices or computers to their TV.

The relatively low numbers of BTV loyalists speak to the expanded viewing options such as industry-sanctioned streaming services, which do not require cutting the cord/antenna and/or getting rid of the living room TV set. BTV may be costly in locations where cable+ services are required, but for most of the participants, HTV assemblers included, the cost was outweighed by the convenience and range of available programming. The inclusion of DVRs with the higher-tier cable+ packages has further enhanced viewer control over the broadcast flow by blurring the boundary between live and time-shifted viewing.

It would also be a mistake to assume that the bulk of HTV assemblers are in transition, that is, on their way to becoming IPTV trendsetters; many retain DTT or cable+ services and watch some broadcast television because of the interests of other members of the household. Thus, investments in intra-assemblages cannot be explained solely by technological trends, convenience, personal choice, or even a politics of resistance. In the next two chapters, I will elaborate on decisions about delivery services, screens, devices, and modes of viewing as they pertain to both domestic and affective relations.

Notes

1. See Bury and Li (2015) for a more detailed discussion of the statistical analysis.
2. Standardized beta = -0.166, $p < 0.001$.

3. This percentage does not include the use of a mobile device as a second screen while watching television content on a TV set.
4. Standardized beta = 0.208, $p < 0.001$.
5. To comply with recommendations made by the Research Ethics Board at my university, I did not distinguish between unauthorized streaming and downloading activity in the questionnaire.
6. Downloading: standardized beta = −0.166, $p < 0.001$; streaming (third party): standardized beta = −0.106, $p < 0.05$; streaming (cable and network): standardized beta = 0.049, $p > 0.05$.
7. Louise's mention of time-shifting above was made in reference to streamed, not recorded content.
8. As was the case with the survey I was careful not to ask direct questions in the interview about unauthorized downloading. A few participants still chose to talk directly about using BitTorrent and the rest made indirect references.

References

Bury, R., & Li, J. (2015). Is it live or is it timeshifted, streamed or downloaded? Watching television in the era of multiple screens. *New Media & Society, 17*(4), 592–610. doi: 10.1177/1461444813508368

Haraway, D. (1988). Situated knowledges: The science question in feminism and the privilege of partial perspective. *Feminist Studies, 14*(3), 575–599.

Ihde, D. (1990). *Technology and the lifeworld: From garden to earth.* Bloomington: Indiana University Press.

· 3 ·

TELEVISION 2.0 AND EVERYDAY LIFE

Although television has reassembled itself several times since its takeover of the radio assemblage, it has remained true to its radio roots—a domestic technology. Television, states Silverstone, "is watched at home. Ignored at home. Discussed at home. Watched in private and with members of the family and friends" (1994, p. 24). This chapter focuses on the ways in which multiscreen and multimodal viewing, whether done by BTV loyalists, IPTV trendsetters, or HTV assemblers, is imbricated with domestic relations and the routines of daily life. The home is not just a physical space but a cultural construct with a history that dates back to industrialization. According to Alan Tomlinson, "the Puritan notion of the home was as a Little Kingdom. The Victorian concept stressed Home as Haven" (1990, p. 48). The contemporary multimedia home, even more so than it was in the early days of analog TV, is a site of a "privatized and atomized leisure and consumer lifestyle" (p. 47).[1] This chapter draws primarily on the social uses typology created by James Lull (1990) to examine the viewing of television as a domestic leisure activity. While television may be watched alone, it still takes place within a set of social and spatial relations. Lull divides his typology into two major categories: structural and relational. In the first half of this chapter, I will present data samples from the interviews to focus on the environmental and regulative functions of television, that is, the use of television as background and to punctuate time and activity. In the second half, I will focus on television's role in facilitating engagement with other members of the household, with a focus on social viewing, or *co-viewing* as the practice is referred to by some communications scholars and the audience measurement industry. As I will demonstrate, viewing patterns associated with Television 2.0 are very much a part of the routines of everyday life.

Structuring the Everyday

Television as Background

According to Lull, television can serve as "a flow of constant background noise" (1990, p. 34). One third of the participants made reference to background viewing, almost exclusively in relation to live viewing:

> The only live television I will watch is if I am doing something else around the house and then I use the television very much as people used to use radios. It's just on in the background. So that will be the only live [viewing]. So that's like in the morning if I am busy getting ready for the day or whatever. (Anne E.)

> When we do turn on the free-to-air TV typically what would be happening is I would turn it on maybe not quite first thing in the morning but fairly early on. My eldest daughter who is three and three-quarters watches PBS kids' programming for about an hour or two. It will just be on while we are kind of doing whatever we are doing in the living room. (Jake)

Several other participants including Notesofwhimsey and Idoru also reported that most of their live viewing had become background; the rest was time-shifted or done online. Ten specifically used the term "noise," which along with Anne's radio analogy suggests that the some background viewing is better described as listening.

> I like to just have the TV on. I live by myself and we are not allowed pets or anything so it's nice to just have it on for background noise. So yeah, I will just flip around until I find something either that I haven't seen or that I can stand to watch for the thousandth time and just sort of leave that on in the background. ... When I am flipping around I am usually hoping that I will find either a rerun of *Law and Order: SVU*, or *House* somewhere because they seem to be on all the time. But if I have my preference then I would probably rather watch like a science fiction or a fantasy show or one of the hour-long dramas. (Bunny)

Bunny's mention of living alone fits with Lull's (1990) claim that companionship is one aspect of television's environmental function. Her comments also highlight the importance of finding the "Goldilocks zone," one which allows programming to successfully function as background: either previously unviewed episodes in a familiar series in syndicated rerun or ones not seen so many times as to become boring. Amy made a similar point about reruns for this function: "I think it's more relaxing because you know it's going to be good because you already saw it and you know what you can expect from it." Margene watched *Law and Order* for a similar reason:

> There are things that are on at times when I am just looking to watch *Law and Order*. Sometimes I just want that background more than I want to pay attention and so I would otherwise probably be missing things that I might *want* to see in order to have that sort of background noise. (Margene)

Procedural dramas also have a highly formulaic narrative structure, which can make them easier to follow without paying close attention. For their part, Joan, Jayne, Rain, Buffy, and Diva mentioned tuning to TLC, the Food Network or HGTV, American specialty networks that air lifestyle and reality programming that are also highly formulaic. Lull makes the case that background viewing can shift to attentive viewing depending on the moment and context, as happened to Margene with *Say Yes to the Dress*:

> This ridiculous TLC half-hour reality show about people shopping for wedding dresses. It's fantastic background noise but then in the past year I actually was shopping for a wedding dress so I started watching it. I expect that that will stop. (Margene)

A significant feature of background reception, as suggested by the samples above, is its ties to simultaneous engagement with other household activities. According to one study, almost two-thirds of television viewing is incorporated with another activity (Kubey & Csikszentmihalyi, referenced in Gauntlett & Hill, 1999). It is not clear whether Knitmeapony was using her laptop for work or leisure but Karen and Nem both specified that they liked to have the TV on while doing academic work:

> Often during the daytime I will, I work from home quite a bit now that I am working on my dissertation so I will turn it on to something like *Supernatural* reruns on TNT and then when that's over, flip around and try to find something else. I spend quite a bit of time on the History channel and on BBC America. That's not scheduled, it's just sort of background noise while I'm working. (Karen)

> I will surf channels and then put *Road Wars* and *Motorway Cops* on while I am working because it's stuff I don't have to concentrate on. (Nem)

A number of participants who did background viewing did so in relation to daily routines. Both Anne E. and Jake mention the use of television in the morning, in her case while getting ready for work and in his while minding his children during the day. Others talked about viewing television content while engaging in household tasks such as making dinner (Freda, Tasha, and Margene). Heresluck mentioned she folded laundry, although her viewing was not necessarily distracted. Karen elaborated on viewing in relation to another activity rather than a household task:

> And then I went through a phase where I was watching a lot of design shows on HGTV so I would just sort of turn that on and … sort of leave that on kind of as background noise while I was working or doing other things. I do a lot of knitting. I started knitting to have something to do while I was watching TV. So now when I knit, I always have to have the TV on pretty much. So it started off as I want to do something while I am watching TV but now I knit quite a bit so I watch more TV in order to knit. (Karen)

This quotation provides an example of the ways in which viewing and certain leisure activities can become completely intertwined in everyday life.

Live viewing is not the only mode that enables background viewing: Knitmeapony, for example, used the DVD player on her main computer:

> It's got this super awesome DVD drive and all that good stuff in it and it is connected to the big monitor. So I can see that monitor from the couch. I will be on my laptop doing one thing and be playing a DVD on the other. I guess I almost use it like a TV sometimes. Then I will leave it on in the background and half watch like a whole disc worth of whatever. (Knitmeapony)

Similarly, Anna B. mentioned putting in a DVD from a box set of *Buffy the Vampire Slayer*, one of her favorite series, while "cleaning or doing a task; it's good background noise." GSM downloaded programming in the morning from the previous evening to watch on his computer: "If I am sitting at the computer I can run it in a window." In light of the above, the practice of background viewing still relies primarily on live TV although other modes are mentioned. Indeed the ability to turn on the television and engage in other tasks and activities is one of the primary reasons that most people are not cord cutters.

Making Time for TV

Television viewing of course is often a standalone leisure activity, not just a backdrop for other activities. As such it may be planned or spontaneous. Gauntlett and Hill (1999) found that sixty-one percent of their research participants did not organize their schedules around specific television programs, twenty-five percent did so sometimes and only ten percent did so with regularity. The results from the TV2.0 survey were the opposite: only thirty percent of the respondents did unplanned, unscheduled viewing of live television that may or may not have involved channel surfing. Inferential statistical analysis (ANOVA) found significant differences across regions, with thirty-eight percent of those living in Australia/New Zealand doing the most and those in Canada doing the least (twenty-five percent). Moreover, just over a third of the male respondents did unplanned viewing and/or channel surfing compared to twenty-seven percent of the female respondents. Whether planned or unplanned, the interview data clearly demonstrated the ways in which leisure viewing was incorporated into personal and family schedules. Mary, for example, talked about watching *Headline News* on CNN in the morning before work and then reruns in the evening after dinner before she went to bed at 7 pm in anticipation of an early start and long commute in the morning. Fifteen percent of Gauntlett and Hill's (1999) respondents said that they often watched TV while eating a meal. Several TV 2.0 participants mentioned doing the same. Vera, for example, found that "police documentaries are very handy because they are on around dinner time." William and Khal mentioned watching the news while eating dinner while Daisy and her husband (both retired) enjoyed watching reruns of *The Big Bang Theory*. Television is also part of the routines of shift workers as Buffy once was: "I worked at catering so I would get home really late and it was *Sex and the City* and *Queer as Folk* back to back."

Channel surfing was mentioned by a number of participants who wanted to watch TV to relax but did not have a particular program in mind:

> I am not the kind of person who makes an appointment to watch let's say something. There is a great element of serendipity in a sense. So when I happen to switch on the TV I will just channel surf until I find something that might be interesting. (Glocal)

As for Jake, once his wife came home and took over child care, he would watch TV "to decompress":

> I channel surf between syndicated repeats of *Two and a Half Men* or *Jeopardy* or pretty much whatever is on 6:00 that kind of catches my interest. Occasionally I will turn to someplace like some of the news channels but because I get a lot of my news on the internet and a lot of that is redundant information so I tend to gravitate more towards entertainment. (Jake)

Kate, Farah, and Khal talked about using the electronic on-screen guides provided by the cable+ providers. Farah aptly described the use of the guide function as "menu surfing" given that many systems include a pop-up screen of the program being broadcast when the channel is highlighted.

> So what we will use is we will bring up the guide, the kind of digital guide that they have as part of their interface. We will look at all the channels that we have access to and see if there is anything worth watching. Generally there's not, I have to say; which is a sad thing. But occasionally there is and sometimes there might be a show that is a good kind of family show. (Khal)

Several participants described their selection sequence:

> I usually start with the commercial networks and then I check out the BBC channels and then I will check out the German and Flemish channels. I probably save the Science channels for the last. (Vera)

> So I usually do like CBC. I will do the local CBC and then [CBC Newsworld]. I will look to see if Seinfeld is on because Seinfeld is always on. HBO, I will always see what's going on on HBO. TVtropolis and stuff. Yeah, certain channels where I tend to watch a lot of their programs I will flip around to kind of see what's going on there. (Tasha)

As these samples indicate, specialty cable channels were often the focus of these directed searches. Among the US channels mentioned were The Animal Planet, The Discovery Channel, the Food Network, the History Channel, National Geographic Channel, and TLC. The next set of samples demonstrates that searches for content for leisure viewing were not limited to live TV:

> If I am home for dinner and I just finished cooking dinner then I will just sit and watch whatever is on if I don't have anything on my DVR. (Rain)

> If there is nothing on TV as in scheduled TV, I maybe switch the iPlayer on and just have a browse through what's on the iPlayer that night and if I really want to watch some TV just to chill out I will look through maybe some documentaries or arts-type programming though the iPlayer. (Will)

While Rain started by searching her recordings, Will started with the broadcast schedule and then switched to his BBC streaming service. Although she loved Netflix's extensive archive, Jayne complained about the lack of updates to the menus and content:

> So by the time we get to the end of the month and you are sort of sick of looking at the same 50 movies beyond all the scrolling capabilities or searching capabilities that you have, we tend to start flipping through the television channels. Inevitably, we end up back watching the Food Network! (Jayne)

Like any leisure activity, watching TV is also constrained by professional, personal, and familial responsibilities. Time-shifting was described in the previous chapter as enabling viewer-centered flow. The following samples provide details of the ways in which it was used by some to reorganize the broadcast schedule to fit with their own:

> I leave at 6:00 to get to work which means I am normally in bed by 10:00. Say there is a 2-hour crime drama on BBC One starting at 9:00; there is no way I am staying up until 11:00 to watch the end of it. So I record it and then watch it at another time (Nem)

> I am quite busy all week and I am usually doing stuff and I get up pretty early to come to school so I can't watch a lot of my favourite programming when it airs live. So I have a nice DVR and I watch it the next day. (Rene)

In Suzie's case, her viewing had to be organized around her son's bedtime:

> I pretty much record everything that I watch. I have a 10 year old son and he goes to bed around 9:00 and that's when either an 8:00 show has just finished or a 9:00 show is starting and I just prefer to record it and then be able to, even start a 9:00 show at 9:30 or something like that after he has gone to bed. (Suzie)

Others gave priority to other leisure activities:

> If I for example am home on a Thursday night, then I will watch the Thursday night line-up on NBC. If I'm not home for some reason, I have competing obligations, I am not able to be here then I would then view that Thursday night line-up on NBC over the weekend—the subsequent weekend. So I may watch one hour of the programming on a Saturday if I've got a little bit of time in the afternoon and then on Sunday I would pick up the rest of it. (Kim)

Programming was thus shifted by as little as half an hour from the time of broadcast to the following weekend depending on how it best fit with the participants' schedules and lifestyles.

While not all leisure viewing involved live or time-shifted viewing, for the majority of participants, it involved a TV set. This choice makes sense when we consider that televisions and computers have traditionally been associated with different spaces of the home: living rooms and bedrooms versus offices and dens. Comfortable seating is emphasized in the next set of samples:

> I can sit on my couch as opposed to sitting in my desk chair and do other stuff like fold laundry while I watch TV or just hang out on my much more comfortable couch with a cat on my lap or whatever. So yeah, I am at the computer enough as it is. I don't feel inclined to spend all my TV watching time also in this chair. ... Comfort is part of that but even just aside from that, just being in a different space is nice for me. (Heresluck)

> Yeah, I prefer to watch on a television screen when I am at home because it's a more social form because it is more comfortable. Because of the way that the room is set up I can sort of relax on the couch and watch. In general if I am watching it on a computer screen I am multitasking with it. If it's on the computer screen I will also being doing email, I will be writing a chapter, I will be doing other work or personal related things. (William)

These participants contrasted the comfort of viewing in a living room on a sofa with viewing on their computers. The desk computer was deemed unsuitable because it was not only located in a work space but in William's case it was used for a series of other work and personal activities that detracted from relaxation. Tarsus had a computer and a TV set in the living room but sometimes watched programs on the first because she could not be bothered to get up from her desk.

Responses were more mixed about viewing on a laptop. As noted in Chapter 2, Jake preferred to do his leisure viewing lying on the sofa with his laptop. Stevie held the opposite view:

> I would much rather watch that on a TV than on my laptop just because I don't really like sitting with my laptop on the couch and then sitting at a table with my laptop or at the desk doesn't feel as comfortable I suppose. So it's kind of like you know, if I am going to relax and watch something, it's going to be on the TV. (Stevie)

The bedroom was mentioned by a few participants as a space for leisure viewing, including Douglas: "Sometimes I want to lay in bed and watch something." For those without a TV in the bedroom, the laptop held particular appeal: Anna B. and her husband liked to use their laptop to watch TV in bed, even though they had a larger computer monitor set up in the living room instead of a TV: "I think watching TV is quite an intimate thing. For me, anyway. And I quite like being up close to the screen."

> We only have one TV. If I am feeling particularly lazy and I want to just stay in bed I will use just my laptop in the bedroom but mostly we connect a laptop to the TV because we have a big flat screen TV so we can get a better picture. (Joan)

> You can sit with it in your bed and basically just, you know, snuggle up and watch and then fall asleep. (Lisa)

The participants also held mixed opinions as to the use of mobile devices in relation to comfort and relaxation.

> I do have Netflix and I have watched Netflix via the Wii. I am kind of starting to experiment with my iPad and with other equipment that I own. But it's still sort of the same kind of computer issue of I would rather lie back and watch television if I am going to watch television. (Aimee)

That said, Aimee did note that the iPad was "bed friendly in a way that a laptop is not." Ellen also felt that an iPad was a better option than purchasing a new TV set for the bedroom:

> When I started getting hooked on all these Turkish shows and realized that my television in the bedroom was completely ancient I looked at the cost of buying a new television versus the cost of buying an iPad and thought well I can do the exact same thing on the iPad but I will have all these additional benefits as well. So I justified getting an iPad as a portable television. And then once I got it I understood why people were screaming about it because it is so much more than a portable television. But it definitely does the portable television thing down to a T. (Ellen)

Douglas, however, did not see the point of watching TV on an iPad while at home:

> Like you can get apps for the iPad and you can get stream or watch TV and all of those things on your iPad. But you need to be at home connected to your Wi-Fi and if I am at home then I am going to watch it on TV. So I am not understanding exactly what point those serve. Someone said, "What if your wife is watching TV?" I am saying then I have another huge TV in the bedroom. "But what if somebody is watching that or you are recording two shows?" Okay, then I have a TV in the guest room and I have a PVR. So I mean, there is no reason for me to need to watch TV on an iPad or a mobile device while I am at home. (Douglas)

Similarly, smart phones were not seen as suitable screens for home leisure viewing.

> If we were camping and stuck in a tent somewhere then maybe the mobile phone would be worthwhile. But if we are at home, why pull out your phone when you can just sit down and watch the TV. (Camden)

These findings are in line with those of the survey respondents. Of the four percent who had watched television content on a mobile device, only ten percent had done so at home. Taken together, these findings suggest that television may be untethered but is not quite as "site unspecific" within the home as Dawson (2007) contends. That said, the TV 2.0 survey results suggest that mobile devices are enabling television "on the go": one quarter of the mobile viewers had engaged in mobile viewing while traveling. Commuting was offered as another reason. Willow, for example, had an iPhone and downloaded content in advance for her two-hour daily commute by bus. In the context of leisure viewing in the home, phones and tablets are more likely to be used as a second screen. The TV 2.0 survey only asked questions about fan-related uses of a second screen, the results of which will be discussed in Chapter 5. However, it is worth noting here that the Pew Research Center found that fifty-eight percent of Americans had used their smart phone as a "distraction device" (seventy-three percent of those aged eighteen to twenty-four) during commercial breaks (Smith, 2012, p. 5). Sherryl Wilson found that the majority of her research participants used social media "as an adjunct to television viewing when not fully immersed in a TV show" (2016, p. 184). Similarly, Evelien D'heer and Cedric Courtois (2016) found that their participants regularly engaged in checking email and Facebook as well as web browsing on mobile devices when not in focused mode of viewing mode. The only example of such use from the interview data did not involve a second screen but rather a second window: GSM commented that he liked to check Facebook and email while watching his downloaded content. To sum up, leisure viewing has not lost its centrality to everyday life, although it is clear that the live mode has lost some ground to the other modes, including DVD playback. Similarly, the TV remains the preferred screen due to its central location and large size.

The Family That Views Together ...

> While we are kind of not traditional in the sense of how we get content, we are kind of traditional in how we watch it. So we will sit in the family room or what have you and watch the television.
>
> —Khal

Since VCRs and multiple household sets became common in the 1980s, families have not been limited to watching television together in the living room.

As a number of the samples presented in the previous section suggest, much of the leisure viewing that the participants did as part of their daily routines was also done with other members of the household: spouses/partners, children/parents, and/or roommates:

> We tape probably 6 or 7 shows every week. I am making this up but it's approximately true; probably 12 hours total, most of which we will watch an hour a night while eating dinner, maybe 2 hours a night while eating dinner and I have some physical therapy I have to do. Not every night. I didn't watch any television last night. Typically Friday nights we will watch a bunch, whatever we haven't seen through the week. (Willow)

Some participants made it clear that they always watched TV with their respective partners: Kate, for example, said about herself and her husband, "We always share. We watch the same things." Similarly Robert stated that he and his partner didn't have any conflicts over programming selection, "I am almost always watching television with the other person so there are no real issues." Jake co-viewed with his wife: "Anything that she wants to watch is something I would typically want to watch as well, unless it's sports." While there were sometimes differing levels of investment in a particular program, co-viewing took precedence, especially among couples. As Lull argues, television has the "potential as a resource for the construction of desired opportunities for interpersonal contact" (1990, p. 38). For example, Corsac stated, "*Grey's Anatomy*—that's more my girlfriend but I watch it with her most of the time."

> I think this summer we've got the dance reality, *So You Think You Can Dance*. That's one that my wife likes to watch so we do that one. In fact, actually I think that might be the only thing on the schedule right now. In the fall and winter we actually do a lot of what's on broadcast TV. So we do *Glee*, just the *Law and Order* shows we like. I like *Community*. My wife kind of likes *The Biggest Loser* so that one is usually on the schedule. We do a few cable shows. We both really like *Mad Men* and *Breaking Bad* is one that I watch. (Camden)

> I'd still say it's probably say of that 70%, 85% is me and my husband. No wait—60% is me and my husband, 20% is me by myself, and then 25% [when] we'd have people over. (George F.)

Mary and her husband had what she described as "very, very different television habits." Yet they still did much of their viewing together: "I have learned to tolerate watching shows about snakes eating live little animals and he has learned to tolerate watching shows with fake dead people and autopsies."

Content plays a bigger role in co-viewing situations that involve more members of the household:

> Usually if it is like the sitcom or a drama series like *House* or a sitcom like *The Big Bang Theory*, it would be more likely to be something that we would all watch together. Like other people in the home. But other things that are more like documentaries and things, those tend to be more individualized unless it's something that happens to be something we are all interested in. (Brooke)

> My viewing was then very much circled around what everyone wanted to watch which would have been different at different times. My housemates did watch *Sex and the City* so I kept watching it but we also liked to watch things that the whole group could kind of get into. So we watched a lot of things like *Jeopardy* and *Survivor*. So some of the reality TV. *American Idol* I guess. Those kinds of like very accessible shows to everyone that you could get into and watch as a group. (Buffy)

Popular prime-time dramas, sitcoms, or reality series were perceived as the most suitable content for social viewing.

Of course not all couples or family members want to do all their viewing together, and thus viewing schedules have to be negotiated depending on the household configuration:

> For example, I like the *Real Housewives* franchise on Bravo. I tend to tape those because the other person in my household doesn't like them. But I usually watch them early the next morning. Do you know what I mean? So I tape it that night and I either watch it like later that night after she goes to bed or because she leaves for work earlier than me so I will watch it right early in the morning. (Farah)

> We watch *Harry's Law* because my husband doesn't like *Castle*. We'd PVR'd *Castle* and I would watch it afterwards and then during the ads we would watch *Hawaii Five-0*. (Notesofwhimsey)

As illustrated by the above samples, the participants had to compromise. Time-shifting is one means of resolving a potential conflict. Separate TVs in different spaces were also mentioned for those who wished to watch different programming at the same time:

> We have four TVs but only one has the DVR. Like with Verizon you can have whatever it's called, multiroom DVR or something like that. We don't have that. The idea there is if you are in the bedroom you could still watch something that was recorded. Like ours is on our main family room TV. So it does kind of create some problems because my husband likes to use that family room TV a lot. So I guess that's when I might be watching something not on the DVR is because he has that TV so I don't

have access to any of the shows on the DVR. So then I will go to the bedroom or something and watch HGTV or something. (Suzie)

In Suzie's case recorded content was valued more highly than live TV. In Mary's household, having a second DVR (or one with a multiroom function) might mean no longer having to watch her husband's nature programs:

Actually, I have a TV in my little sitting room/office. He has a TV in his office. There is a TV in the bedroom and a TV in the main room. I mostly watch TV back here on my television. When we watch TV together or when I decide to watch something on the DVR I have to go out to the main room because that's where the DVR shows are available. He can watch DVR shows in his office. I would like to change the setup. I would like to actually have 2 DVR's. But that's not in the cards just yet. I am working on it. So he can have his DVR and I can have my DVR. (Mary)

The use of computers also accommodated differing tastes or schedules.

Until recently I shared a one-bedroom apartment with my partner who is a game studies scholar. He was often using the TV and consoles and then I would often watch something on my laptop in another room. (Margene)

We've got a living room with one non-digital TV. We don't have cable so we are using UK terrestrial channels. We do have a DVD player as well so all of my watching on the television is done usually in company with family. (Zee)

Zee watched the rest of her programming on her laptop in her room. Domestic relations also determined Freda's choice of screen:

I do watch television on my laptop and on the TV. It's kind of 50/50. It's the same programs but which screen I put it on is kind of, it depends on the context. ... If my housemates are watching the news then I will just watch it on my laptop but if no one is using the television or sometimes my housemate and I will get together and watch something that I have "procured" [online]. (Freda)

The next two samples would seem to support Jason Jacobs' claim that "digital television threatens the universal experience of television's social function" (2011, p. 267):

There are 5 people living in the house. Myself and my husband and three out of four children. Up until a month ago there were still the same number but it was a different child living at home. One moved out and one moved back in. And mostly I would have to say since we now have a wide screen TV, we probably watch less television now than we *ever* have in our family's history because we only one TV in the house but we have 8 computers. So most watching by the children who are between the ages of 15 and 24, most of that is done on computer. (Notesofwhimsey)

Lately I just watch it on the laptop. It's easier that way. It's just that I have been living with my sister and my brother-in-law for the past 8 months maybe and we have a TV set in the living room. Like I said, I have a nephew and he likes to watch TV and my computer is just easier. I watch it in my room. It's just there. I don't have to record a DVD and go over there to the TV set. (M)

Although there were exceptions (Anna B. quoted earlier being one), most participants did not think that computers and mobile devices were suitable for co-viewing. That said, the small screen problem can be resolved by hooking up computers and laptops to TV sets. Courtney, a cord cutter with no TV set, stated that her laptop suited her needs for viewing alone. Then she added, "Now I am possibly going to get into another relationship so it makes sense to have a better viewing solution than a laptop in case like, somebody else is over." Her solution was to connect her laptop to a projector. Revan and his wife often watched online programming in their respective offices, but if they decided to view online content together, their living room TV was also connected to one of the PCs. Another solution afforded by IPTV to support social viewing among those with different viewing tastes is the simultaneous use of multiple screens in the same viewing space. George F. stated, "We usually have our laptops with us as well. So we might be watching something or somebody might be gaming and somebody might be watching something."

A final aspect of co-viewing to consider is viewing as a social activity with friends rather than with other members of the household. George F. was quoted earlier in the chapter as saying that one quarter of her viewing was done with friends who came over to visit. Rene recalled having a "TV night where a bunch of us would come over to watch Sci-Fi Friday. We would cook for each other and do a communal Sci-Fi television." Similarly Lisa recalled social viewing with other nursing students: "Thursday would be our group days so we would watch ER and *Friends*." For Freda co-viewing *Supernatural* "became like a ritual thing where when I lived in Toronto one of our friends would come over or two of our friends would come over the night it was on and we would always watch it together." Sophie reminisced about watching *So You Think You Can Dance* in a group:

I guess the thing that I sort of miss is getting to watch shows with other people. There are some of the shows, especially some of the reality shows that we used to watch in big groups with friends, that it's easier to do if you have the TV and the cable. (Sophie)

IPTV can also extend the social function of television beyond the confines of the living room:

> We will watch television together over distance and IM each other at the same time. One of my friends doesn't have TiVo or anything like that so she just watches it live and that's it. (Karen)

Karen also mentioned another friend with whom she liked to co-view but who did not live in the same time zone and went to bed early. "If it's not a show that I absolutely have to watch live, I would record it and wait to watch it with her at a later time." To facilitate this practice, a mobile device was sometimes used as a second screen:

> Because I moved and far away from everybody I know—usually I am on Instant Messenger watching the shows with people. ... I usually have 3 or 4 different conversations going if I am watching actually. We call it date night. (Rene)

> Most of the people that I am talking to are also sitting there watching it by themselves and you know, sometimes instead of you trotting through the commute and going across London to my best friend's house, we just turn on Twitter and we just chit chat from our own houses but it still feels like we were sitting together and watching the Oscars together. (Ellen)

While one can do this kind of distance co-viewing with one other person by phone, the practice is clearly facilitated by messaging and social media apps. In short, social viewing is clearly alive and well in the contemporary household, and computers and mobile devices play a role in facilitating sociality. Time-shifting and now streaming and downloading technologies also facilitate individual viewing, resulting in household members no longer having to tolerate differing taste if they chose not to do so.

The More Things Change ...

As I hope to have demonstrated, Television 2.0 remains fully imbricated with the practices and routines of everyday life. First it retains its environmental function from its analog days. A range of participants from BTV loyalists to IPTV trendsetters who still receive a broadcast signal continue to primarily watch live television as background. Moreover, such viewing often involves multitasking in relation to work, domestic chores, and parenting. Mobile devices serving as a second screen are also used in relation to distracted viewing, especially for the purpose of accessing social media during commercial breaks. Television also retains its regulative function to punctuate time as part of daily routines, particularly in relation to getting ready for work and evening meal times. Some of this viewing is organized around specific programming; other times it involves channel surfing. In the context of evening leisure viewing, household members are more likely to engage in time-shifting over other modes, creating their own viewer-centered flow and otherwise integrating television with other aspects of daily life. Such viewing is no longer single screen, although the TV set is still given preference given its size and location in the living room. Laptops and tablets, however, serve to untether television, allowing viewing to take place in other domestic spaces, particularly in the bedroom.

As for co-viewing, couples with no children or young children were most likely to watch much of their programming together; families with adult children watched the least amount together. All three modes of viewing are engaged in for this purpose, a finding that disputes the claim that IPTV is curtailing the social function of television. While streamed or downloaded content was sometimes viewed alone on a tablet or laptop, it was also outputted to a large computer monitor, projector, or TV set to facilitate co-viewing.

A theme that runs through this chapter but was not directly addressed was the television content itself and the viewer's relationship to it. Lull's typology downplays the significance of content in relation to both regulative and social functions, and only includes a general statement about television's "timeless environmental function as a source of entertainment for the family" (1990, p. 36). Even when used for background, the TV 2.0 findings suggest a preference for programming that is formulaic or has been viewed previously. Popular prime-time programming operates as a social lubricant that brings members of a household together even if they may be viewing different content on different screens. Moreover, the typology fails to take account of fannish engagement with television texts. The regular "date nights" among friends described in the chapter for example clearly involve collective fannish viewing. At the same time, it is important to remember that sometimes the desire to co-view will trump content, particularly among couples. The next chapter will explore affective relations with the television text and viewing practices that involve a more focused and attentive mode of reception.

Note

1. Although not the focus of the TV 2.0 study, it is important to keep in mind that leisure in the home needs to be understood as imbricated with gender, class, and race. As reported, the majority of the TV 2.0 respondents and participants were middle class and white and thus able to engage in television viewing as a domestic leisure activity.

References

D'heer, E., & Courtois, C. (2016). The changing dynamics of television consumption in the multimedia living room. *Convergence: The International Journal of Research into New Media Technologies, 22*(1), 3–17. doi: 10.1177/1354856514543451

Dawson, M. (2007). Little players, big shows: Format, narration, and style on television's new smaller screens. *Convergence: The International Journal of Research into New Media Technologies, 13*(3), 231–250. doi: 10.1177/1354856507079175

Gauntlett, D., & Hill, A. (1999). *TV living: Television, culture and everyday life.* London and New York: Routledge.

Jacobs, J. (2011). Television interrupted: Pollution or aesthetic? In J. Bennett & N. Strange (Eds.), *Television as digital media* (pp. 255–280). Durham, NC: Duke University Press.

Lull, J. (1990). *Inside family viewing: Ethnographic research on television's audiences.* New York: Routledge.

Silverstone, R. (1994). *Television and everyday life.* New York: Routledge.

Smith, A. (2012). The 'rise' of the connected viewer. Pew Research Center. Retrieved from http://pewinternet.org/Reports/2012/Connected-viewers.aspx

Tomlinson, A. (1990). Home fixtures: Doing-it-yourself in a privatized world. In A. Tomlinson (Ed.), *Consumption, identity, and style: Marketing, meanings, and the packaging of pleasure* (pp. 40–51). London and New York: Routledge.

Wilson, S. (2016). In the living room: Second screens and TV audiences. *Television & New Media, 17*(2), 174 – 191. doi: 10.1177/1527476415593348

· 4 ·

AFFECT AND THE TELEVISION TEXT

This chapter focuses on affective relations and fannish viewing practices as they pertain to serial narratives. According to Fiske, all genres of popular TV series evoke a sense of "nowness" and deny closure:

> However completely the plot of one episode is closed off, the situation never is. All television series have some characteristics of serials—the relationships between the characters and the possibilities of the basic situation are never completed or resolved, abut remain open, reverberating, and ready for reactivation next week. ... Television's tension between the forces of closure and of openness, between authorial and viewer authority, still remains central to the textual experience they offer. (1989, p. 69)

As such they encourage viewers to actively engage with them on an ongoing basis, thus holding out an offer of commitment, albeit at a pace determined by the broadcasting network. When one commits to regular viewing, one is effectively entering into an affective relationship with the text. All but ten percent of the TV 2.0 survey respondents (ninety-two percent of women, eighty-four percent of men) were committed viewers, having watched the majority of episodes of at least one series over the previous twelve-month period: the average number of series watched during this time period was seven. Eighty percent had watched all the seasons of the first of two series that they were asked to identify by name. The list of titles based on the responses to this question was dominated by dramas, followed by sitcoms and reality series. The list included both current series and some that had long since completed their runs. As expected, all but one of the series named were English language (the exception was the German soap opera *Alles was zählt* (*All that matters*), the vast majority of which were American productions. That said, a dozen or so British series were named, along with a few international coproductions, such as *The Tutors*, *Downton Abbey*, and *Orphan Black*. In the interviews *Doctor Who* and *Lost* (the final season aired during the survey period) were mentioned by over half the participants. *Star Trek* and The *X-Files* were mentioned by at least one quarter along with popular series current at the time of the interviews: *Dexter*, *Game of Thrones* (which only premiered after the survey closed in April 2011), *Grey's Anatomy*, *Mad Men*, and *True Blood*. These findings confirm the global reach of American serial television and its enduring popularity with both American and international viewers. In the pages that follow, I will present samples from the interview data to further explore the formations of and variations in these long-term, sometimes "long-distance," multiserial, affective relationships associated with Television 2.0. I will also discuss the viewing practices that they engender, including anticipatory, catch-up, repeat, and so-called binge viewing as well as the related practice of DVD collecting.

Making a Commitment

In order to get a sense of when and how one becomes a fan of a series, the TV 2.0 survey asked respondents at what point they became committed viewers of the two series they named: For the first, nearly half (forty-seven percent) reported during the first season. Some of the interview participants made a point of checking out the new network shows. Daisy, for example, said, "My life gets very crowded in September when the new shows come on while I try to sort out what I am liking and don't like." Lisa made a similar point:

> Yeah, basically I think everything that is like new I kind of try it. ... *The Event*, I tried that. Yeah, so everything basically I try it once and then if I like it I stick with it, if I don't I just leave it. (Lisa)

Rene's decision to try *Sherlock* was based on the reputation of the showrunner: "Being a fan of Steven Moffat's writing I figured it would be pretty brilliant and of course, it was pretty brilliant." Lauchita was anticipating *Mad Men* after reading articles in the *New York Times* and *The Guardian*:

> Then one day my husband came and said they aired the show, that it's perfect. I started watching it and in the first episode I said, "This is very boring. I am never going to like it." But I kept watching it because I couldn't believe the perfect art direction and the clothes. Then in the second episode I was like a total ally. (Lauchita)

Not everyone becomes a fan "at first sight." Bunny, for example, began viewing the anime series *Gundam Wing* as part of her daily routine.

> [It] came on about the time I came home from work and I would just turn it on and kind of got hooked on it that way and from there, expanded out to just a whole bunch of other types of shows. (Bunny)

This pattern fits with Jenkins' (1992) finding that becoming a fan can take place over an extended period of time. Jenkins (1992) also makes the case that reruns of series, including those that are no longer in production, create opportunities to develop an affective relationship. Thirty-two percent of the respondents reporting becoming a committed viewer during a subsequent season and ten percent after the series had completed its original broadcast run. The following participants talked about becoming a fan later in the run and then catching up if the series was still airing new episodes:

> I didn't start watching [*Lost*] until the first season started on reruns. My sister was like, "You have to watch this show. It's about a plane crash on an island." And I was like, "I don't know if really want to watch a show about a plane crash" but finally when I started the reruns of the first season I got sucked in and have been completely addicted to it ever since. (Suzie)

> Some friends actually kept recommending it to me and for some particular reason I started watching *Lost* after, basically it was the end of the final season. So I was seriously out of date! One reason I started watching was precisely because it felt to me that I was losing touch with what was basically popular culture. Everybody seemed to know all about it and I didn't know anything about it at all. So at some point I felt a need. To me it was to down to a sense of duty! It was like a must do thing! (Glocal)

Both quotations also gesture to the value of opinion of others in trying out a series, in the form of direct word of mouth or based on *popular cultural capital* (Fiske, 1989). The TV 2.0 survey found that fifty-three percent of the respondents felt that they had influenced others, and forty-three percent said that they were not sure but they had told others about the series that they named. While just under half also said that they had been influenced by others to become a fan, forty-one percent said they had not.

While Suzie caught up through network reruns, others relied on DVDs or streamed/downloaded content:

> I missed Season 1 of *Breaking Bad*. I caught up on that after. Actually, I guess the same thing with *Dexter* too. Season 1 was done and I bought the DVD box set when it came out and I have been watching it since then. (Douglas)

> I will hear some people talking about a show and it's Season 2 or Season 3 and I think, oh maybe I want to watch that. So I get the DVD in Netflix and watch the first few episodes and at that point say either, "Yes this is great! I want to keep going. I am going to buy the DVD's" or alternatively, "I don't understand what you people see in this show at all. I am officially crossing this off my to-watch list and moving on now." (Heresluck)

> I had to catch up on the first two seasons [of *Mad Men*]. There was obviously a lot of buzz but, as you can kind of get a sense of, I watch a lot of TV already so it takes a lot for me to commit to a new one. So I went back and watched the first two [online] and then got into it starting with the 3rd season. And then became, as I often do, very obsessed with it. (Buffy)

A similar pattern emerged for series that had completed their runs:

> I was never a fan of *House* and one of my friends kept going, "You have to go watch *House*. You have to go watch *House*." So when I sprung for Hulu Plus a couple of months ago I am like, okay all of *House* is on Hulu Plus. I will watch Season 1 while I am doing other stuff and if I get invested enough that I start neglecting what I am actually doing, I will watch the rest of the series. And I did. (Knitmeapony)

> Suddenly people start talking about something/a show and they will say, "Oh, man! This was such a great show." And they will start talking about it and I am like, well I never saw that. So you go hunting around and yeah, you can find them online. You can sometimes find these older shows online. Then I will watch them and like, "Wow! This is great!" So yeah, the internet has been wonderful in that respect. Some shows, they never put them out on DVD and there is no other way to watch them simply because they either got cancelled too soon or for whatever reason they feel there is not a big enough fan base to warrant the expense of creating a box set and maybe not enough people will buy it and so they will take a huge loss on it. (Elly)

> I have used Hulu. I don't like it as much as Netflix but yeah, there are some shows on Hulu that aren't on Netflix … like *Queer as Folk* which I never watched the first time around. (Karen)

> I watched *Doctor Who* like the whole new series starting with the ninth doctor. Watched all of that. I watched all of *Dexter* on Netflix and *True Blood*. Pretty much any show either that I missed the first time around or that I wasn't subscribing to really good cable at the time. (Bunny)

Taken together, the samples emphasize the value of streaming services in addition to DVDs in creating opportunities to develop an affective relationship with long running or older series.

Your Affective Intensity May Vary

The following interview data sample provides a useful starting place for the examination of varying levels of intensity in relation to different television texts:

> The *Big Bang Theory*, *Castle*, *Hawaii Five-0*. There are 3 comedies I have started watching lately: *How I Met Your Mother*, *Mike and Molly*, and *Mad Love*. But they are disposable. Tuesday is *Glee*. Wednesday I have got sucked into *Survivor* this year. I don't always. I often tune in in the last 6 shows but this time I got sucked in. Thursday, the new *Big Bang*. There would be *CSI classic*, *Bones*. I have been taping *Grey's Anatomy* but I have fallen on the wayside. I have put that on the old PVR because I have not gotten into it that much. I have been watching *Vampire Diaries*. Fridays, not a heck of a lot on Fridays. Saturdays, the new *Doctor Who*. Sunday, any new *Simpsons*, *True Blood*. There is one called *Lost Girl*. I love vampire stuff. I am just finishing watching the *Buffy* series; it ends on Thursday and then I won't be watching *Buffy* anymore. (Daisy)

Daisy began her list of series in no apparent order, although *Big Bang Theory* was mentioned twice, suggesting it may have been of more importance to her than the others. She then shifted to listing by genre (the three comedies and the reality show), yet her choice of words indicates that these were shows that she was watching regularly but was not invested in emotionally, referring to them as "disposable" or that she was a reluctant viewer (*Survivor*.) She ended by organizing her list around the weekly broadcast schedule. *Grey's Anatomy* was an old favorite that she recorded on her DVR but was no longer watching regularly. Finally, she identified a fondness for a particular genre. Her comment about *Buffy: The Vampire Slayer*, however, suggests that she did not see the series when it originally aired and had no ongoing investment in the series beyond viewing in syndication.

A number of participants also described having preferences for particular genres:

> I started watching *The Office* pretty early on and now I watch the pile—the pack of NBC comedies. I don't watch *30 Rock* but I do watch *Parks and Rec* and *Community*. I am pretty into those. (Buffy)

> I would say Sci-Fi. Mainly fantasy ones. There is a new show. I think it's American but I am not even sure, which is called *Camelot*. I *love* Merlin from the BBC. I used to watch *Stargate Atlantis*. So I watch *Stargate Universe* right now. I enjoyed *Battlestar Galactica*. I never watched the first Battlestar thing but the new one which is like the prequel. I don't remember the name. It will come . ... *Caprica*. (Phillipe)

> I am a really big fan of police documentaries. So anything about police work I tend to watch religiously. (Vera)

Willow's favorite genre was science fiction. While she stated that she really enjoyed *Dexter*, it "doesn't resonate with me on those levels that the mythic science fiction stuff does. ... I love *Torchwood* and I love *Doctor Who* but they are both on hiatus. There is no ongoing show that has that level of interest for me right now." Mary was a fan of procedurals but she had stronger feelings about some than others:

> I loved watching *LA Law* when it was on but I was never into fandom. But I have always watched *Law and Order*. *NCIS* caught my attention and then I didn't have any real favourite. ... I liked *Criminal Minds* from the beginning but when Joe Mantegna joined the cast with the character of David Rossi that's when it really became my favourite. I never really liked Jason Gideon that much. I spent too much time wanting to slap him. (Mary)

Mary's last comment demonstrates that casting and character can have an effect on one's level of emotional commitment. Karen began watching *Canada's Next Top Model* because she was a fan of Tricia Helfer, who played a regular character in one of her favorite series, *Battlestar Galactica* (reimagined). In addition to genre and casting, some participants described themselves as fans of networks or just television itself:

> I like *all* of the shows on BBC. I really, really do. I mean, *Doctor Who* is the least of it. *Misfits* is just phenomenal. And right down to the costume dramas. (Knitmeapony)

Running through the samples is a language of devotion as per Vera's quotation above, as well as addiction. Knitmethepony, for example, described the USA network as "one of my latest little addictions. USA is doing awesome stuff: *Burn Notice, Psych, Body of Proof.* I think is what it's called. *In Plain Sight.*" She also mentioned being "addicted" to *Game of Thrones.* George F. said that she could only "dream about being able to watch HBO" as the network was not available in Australia at the time. Lauchita declared herself to be "a total TV freak!"

Just because fans feel a deep emotional connection to a series does not mean they are not invested in *bourgeois aesthetics* (Bourdieu, 1993), that is, a set of discourses which emphasize the quality of the writing, acting, and production of television texts (see Bury, 2005). A number of the series named in the survey and in the interviews could be classified as "quality" drama, a form that Jason Mittell refers to as having *narrative complexity*.

> This model of television storytelling is distinct for its use of narrative complexity as an alternative to the conventional episodic and serial forms that have typified most American television since its inception. We can see such innovative narrative form in popular hits of recent decades from Seinfeld to Lost, West Wing to The X-Files, as well as in critically beloved but ratings-challenged shows like Arrested Development, Veronica Mars, Boomtown, and Firefly. HBO has built its reputation and subscriber base upon narratively complex shows, such as The Sopranos, Six Feet Under, Curb Your Enthusiasm, and The Wire. Clearly, these shows offer an alternative to conventional television narrative. (2006, p. 29)

For example, Tabatha was a fan of the European crime dramas that aired in the UK. Mobilizing a discourse of criticism, she described them as being "so subtle and demanding of one's attention":

> *The Killing* –that's the Danish [version]—was the highlight. It was 9:00—10:30 Saturday night and it was just fantastic. I didn't have to look for my remote control. I just sat and I just revelled in this great drama. (Tabatha)

As happened with Daisy in relation to *Grey's Anatomy*, the next samples demonstrate how an affective intensity can wane over time.

> I've had a change of heart with *House* since I filled out your survey. But *House* was one that no matter what else was on that the time, I watched that live. Anything else got recorded. (Elly)

Buffy went from being a fan to an *anti-fan* (Gray, 2003): "I shudder to hear myself say it now but back when the show as on I was definitely a *Sex and the City* fan." Yet some fans keep watching programs in which they have lost interest for the sociality as Buffy did with *True Blood*:

> I really did like it. I think it has the promise to be amazing then never quite lived up to its promise. But it is really fun to watch with a group because it is so excessive and ridiculous all the time. I have a group that comes over and we watch every week and that makes it really fun. (Buffy)

While the above discussion does not provide a detailed explanation of why affective intensities vary from one series to another over time, it does gesture to some of the elements that provide the "color, tone and texture" (Grossberg, 1992) that make committed viewing more than just a part of the routine of everyday life.

Anticipation: No Longer Keeping Me Waiting

Many fans do not simply view a favorite series as part of their routine but actively organize their schedule around it, eagerly anticipating each new episode (Jenkins, 1992). Traditionally, this practice has required viewing the original broadcast live. Several participants talked about their "mustsee" series:

> I always make a date to watch *White Collar* on USA. I love *White Collar*. *Supernatural*, *House*, and *Castle*. ... *Game of Thrones*, of course and then when *True Blood* starts up I will definitely make a date to watch *True Blood*. (Bunny)

> There are also a couple of shows that I am particularly invested in that I like to see as soon as they come out. That's *Doctor Who* and *Being Human* which isn't airing at the moment but those are my two personal "this is my TV date" that I must watch. (Zee)

> It used to be the rule in the family that 10:00 on Wednesday, everybody had to go away because I had to watch *CSI New York* and they were *not* to interfere with that in anyway unless they were bleeding. (Notesofwhimsey)

As the next sample indicates, a sense of anticipation varies depending on affective intensity:

> I kind of get into things for a while and then I get very excited about them. I mean, I guess I said before that I am fine to wait on the HBO shows and then binge on them

later. But I have a friend that does that with *Mad Men* and I just can't wait. Like it just kills me that I can't talk to her about what happened in last week's episode and how am I going to make it until next week to find out what happened. (Joan)

The stuff I insist on watching live is usually Sci-Fi related stuff because that's mostly what I enjoy in general. So some of the stuff I insist on staying awake for no matter how tired I am would be *Supernatural* … *Stargate* back when it was on. (Rene)

Rene's comment also gestures to the way in which fannish viewing can disrupt everyday routines. Some of the participants who were involved in participatory culture provided another motivation for viewing live:

You don't want to miss an episode because if you go on the internet afterwards you will probably find out what happened anyway. So if you want to stay up with the show and keep it interesting while it's happening you kind of have to watch it while it's airing live. (Anna A.)

Lost was another one that if you don't watch it when it's on or close to the time that it is on, like that night, you are left out of the online conversation the next day. (Joan)

It was the practice of a few of the cord cutters or those with lower-tier cable packages to go to other people's houses to see the original broadcast:

Essentially things that got where I was intensely interested in were over the air and something like *Sex and the City*, *Sopranos*, an exception like that, like a current exception would be like *Mad Men*. That's when you make TV a communal experience and go to somebody's house who does have cable and watch it for that night. (Penguin)

My boyfriend has [HBO] though so when I want to watch for example, *Game of Thrones*, I go to his place. Or at my parents, they have it. I don't. (Idoru)

Given that only one third of the respondents watched live TV (see Chapter 2), the interview participants quoted above were in the minority. Additional findings reveal that just over ten percent of the respondents who named a series that was still in production watched all the episodes live, while twenty percent watched most live. Almost half, however, never watched *any* of the episodes live. Similarly, just ten percent recorded all the new episodes, yet almost two thirds reported never recording any new episodes to watch at a later time. When asked about online viewing, thirty-six percent streamed or downloaded all the new episodes, fourteen percent did so for most, while thirty percent never watched any new episodes online. A quarter of the interview participants made direct or indirect references to using BitTorrent clients.

On the surface, piracy appears to be unrelated to anticipatory viewing. The PricewaterhouseCoopers study on piracy, referenced in Chapter 1, concluded that price was a driving factor, as was the popularity of the practice—the "everybody does it" justification (Bothun & Lieberman, 2011). Cost was indeed a rationale offered by a few of the participants who either were unwilling or unable to pay for cable or for a top-tier cable package that included specialty channels such as HBO. Heresluck, a cord cutter with no signal, downloaded all episodes of the current series of which she was a fan. Yet she also purchased the DVD sets when they came out: "As soon as I have the DVDs the downloads are obsolete. They are lower quality; I can't use them for vidding purposes very well." Heresluck described herself as an "ethical pirate," which fits with Kate MacNeil's findings that acts of piracy can indeed "be informed by a strong moral code" (2016, p. 5). The downloaded copy provides fannish viewing pleasures but lacks value for a fan creator such as Heresluck due to its poor quality.

Jenkins, Ford, and Green offer another perspective on illegal downloading: "piracy is as much a consequence of the market failures of media companies to make content available in a timely and desirable manner as it is a consequence of the moral failure of audience members" (2013, p. 16). Patrick Vonderau makes a similar point about market failure in the Swedish context: "75.7 percent of all regularly watched online content is from the United States, an amount dominated by recently released American television series (53.2 percent)" (2014, p. 105). Recalling the discussion on downloading by region in Chapter 2, the engagement in this practice by those respondents residing in Europe and Australia/New Zealand was higher in comparison to elsewhere. The interview data further support this claim: the participants who viewed pirated content did so primarily because of their affective relationship to the text and a lack of timely access via broadcast television. These participants fell into two groups: those who were fans of American series but lived outside of North America and fans of British series who lived in North America, both of which were affected by "the tyranny of digital distance" (Leaver, 2008).

> We've moved out of our traditional watching live to actually torrenting it so we are up-to-date with the US in *Top Chef Masters*. We couldn't stand it anymore. The wait for it was too long. ... [The same is true of] *The Killing* and *Game of Thrones*, neither of them are on New Zealand television. They are not even being pre-advertised as going to be on New Zealand television! (Khal)

> *Mad Men* here in Argentina got lots of news in the national newspaper and there is no way to see it other than the internet because no one bought it and there is not

AMC here but there is a pretty big crowd of fandom. *Mad Men* is like an extreme case for me because it was like, I knew it aired on Sundays or on Mondays. Then on Mondays I was like desperate and I woke up very early and said, "*Please*, download *Mad Men!!*" (Lauchita)

William was from New Zealand and pointed out that American and British programming was often delayed by six months or more after the date of original broadcast and that less popular series may never be purchased for broadcast. M and Peter P (Brazil), Bella (Norway), and Corsac (France) faced a similar situation; they used BitTorrent clients because they wanted to access new content as soon as possible after the original air date. For those in the UK, the delays were generally shorter, so the decision to wait or not depended on affective intensity. Nem talked about going online for *Bones* and *Lie to Me*. She would have gone online for Season 4 of *Californication* but a friend had promised to give her copies of the DVDs. Although Sky generally aired new episodes only about a week after the original US/Canada airdates, she was not interested in the other programming in the package to justify the expense. Her comment highlights a concerted attempt by the global television market, specifically in relation to the US–UK bidirectional flow, to rectify its previous failures, no doubt due to concerns with piracy. Rene had the following to say about *Doctor Who*, which at one point was airing on BBC America one week after the original broadcast on the BBC:

> We would locate as soon as we could. My roommate and I would locate it via "channels," let's put it that way. And we would watch it that way on her laptop. Then when it came on we would watch it again on BBC America. We would do both! Because we loved it. (Rene)

For some fans, downloading is not only done instead of live viewing but in addition to it. Karen noted that the most recent season of *Doctor Who* aired the same day in the US as in the UK: "that's fabulous and so great for me and actually has cut back on my online viewing." When Rene was asked if she would still search online if the delay was only twenty-four hours, she said that she would just wait. This kind of anticipatory viewing thus can be understood as part of an *impatience economy* (Evans, McDonald, Bae, Ray, & Santos, 2016), with decisions about engagement with either the BTV or IPTV intra-assemblage driven by affect rather than technology.

Once More with Feeling

As I noted in Chapter 1, rewatching is a fannish practice that was first made possible by network and syndication reruns but became viewer centered through the affordances of home recording technology. In the context of the first series named in the survey by the respondents, a sizeable majority did not watch new episodes more than once. Of those who did, fifteen percent did so with DVR/VCR recordings and twenty-eight percent did so online via streaming or downloading. Repeat viewing had uneven take up among the interview participants. Some like Camden and Phillipe explicitly stated that they rarely if ever watched a series more than once. The following comments from those who did engage in this practice confirm Jenkins' (1992) contention that it allows for the accumulation of textual meaning—to quote Douglas, "finding those things that I missed or those little clues."

> There are certain shows which I do that with more often than others and there are certain shows which I sort of expected that I would have done that with and have found that I haven't. But I think I have watched the *West Wing* start to finish at least three times. There are certain shows, especially shows that have perhaps have run for 5–6–7 seasons that I find that when I go back to the start there is another level of appreciation I can take from that. (William)

> In *Leverage* they will often have references to various other pop culture things that eat up on the first time through but when you watch it again you are like: oh wait the names that they used for that alias there, that's a reference to this other show or this other part of you know a book or a comic book or something like that. So that's always fun when you discover things like that. (Sophie)

Rewatching can also reveal flaws that were not apparent at the time of original broadcast:

> There is a group of people right now who are re-watching [*The Sentinel*] episodes on DVD and discussing them online. I am going, "Have you looked at how *bad* his hair was in that episode?" (Mary)

Nostalgia was another reason given for choosing to do a rewatch of a series:

> I've watched a lot of shows and older stuff too from the 90's like *Xena* and *Hercules*. That was fun. *Buffy the Vampire Slayer*. Just kind of nostalgic stuff that I remember from high school. (Bunny)

> We have been kind of slowly working through *X-Files*. Lots of nostalgia viewing on there. (Helen)

> Like *Twin Peaks* I am watching for the thousandth time but typically it's not something that I am like: "Oh, I didn't watch *The Sopranos* this week, I better queue it up." I own *The Sopranos* in its entirety but I haven't watched it in a couple of months probably at least. (Joan)

These last two samples also highlight how the pacing of a rewatch can vary, no doubt dependent on a range of factors including other fannish and leisure viewing as well as the routines of everyday life.

The Collectors

Owning and collecting DVDs is a fannish practice that is closely connected to but cannot be reduced to fannish viewing, whether in relation to catching up with a series or rewatching. The majority of the participants owned at least one series on DVD. As Elly put it, "What a treasure those things are. I love them." Some only bought DVD sets if they expected to engage in repeat viewing:

> I know a lot of people who have a lot more TV DVDs than I do. I tend not to buy things unless I'm really sure that I am going to re-watch them and I tend to buy things that I have watched. So if I haven't seen it, I probably won't want to buy it, if that makes sense. (Stevie)

Rewatching was not a requirement for Willow, who exclaimed, "if I have an emotional commitment to a show, I will buy the DVD even if I never watch them!" Other pleasures specific to collection that do not involve viewing include displaying, contemplation, organizing, lending, and giving (Steirer, 2014). A number of the participants had extensive collections; a few even went to look at their collection during the interview to provide titles:

> Some of the TV shows I really like. Like *Lost, Grey's Anatomy, Alias* (I was a huge fan of *Alias*), *Friends, Sex and the City*. My collection is right in front of me so I am kind of reading them. (M)

> *NCIS, NCIS LA, CSI, CSI New York* if I am getting everything. Let's see; I do have all of *Star Trek Voyager*. All time *Voyager* fan. … I hope I am not forgetting anything. I am going to go look at the book shelf. (Mary)

Heresluck and Julianna took great pride in having built up a collection that enabled them to be lenders:

> But when it comes to certain things and especially television shows I *have* to buy it. I *have* to. Television shows I do not buy used at all. It freaks me out. I am like, I am not going to buy that used. Out of all of my collection, my television collection is my proudest masterpiece. People come to me. They don't rent DVDs; they don't have Netflix. When I lived in the dorms people came to me and rented from me. (Julianna)

Elly, on the other hand, was a borrower:

> I do have a friend who buys DVDs and she will loan them to me. Yeah, that's kind of a cross country affair. She lives on the west coast of the US and I am on the east coast so we ship things back and forth. (Elly)

Given their status as collectables, DVDs, like books, can be signed by the series writers or actors. William, for example, mentioned about getting a *Firefly* DVD signed by Summer Glau (River Tam) at a fan convention. Thus the pleasure of the DVD cannot be understand merely as a playback technology.

On a Binge

"Binge" viewing is a practice that has received a great deal of media attention of late. (Henceforth, I will use the older term "marathon viewing" given the negative connotation of the term, e.g., binge eating and binge drinking). VOD and subscription streaming services have led to a rise in its popularity through easy and convenient access to digital archives of older series. In addition, Netflix broke with the paced delivery of network television by releasing a single season's worth of new episodes from an original series at once, an industry shift which serves to legitimate and normalize this practice. Its roots can be traced back to syndication in which the broadcast schedule is intentionally accelerated, through the airing of reruns daily rather than weekly. Moreover, networks have a long tradition of holding marathons for special occasions: for example, the Space network (Canada) and BBC America held a marathon of the original *Star Trek* series to commemorate its 50th anniversary in 2016. The US specialty networks often hold marathons in advance of a new season: AMC for example has done so for *Breaking Bad*, *Mad Men*, and *The Walking Dead*.

What constitutes a televisual binge has never been clarified, leaving Marike Jenner to conclude that it is "defined through highly individualized terms and practices" (2016, p. 265). The interview data revealed several reasons and contexts for this practice, all of which have been discussed previously in relation to other types of fannish engagement. The first involves a rewatch marathon among groups of fans, as described by Jenkins (1992) (see Chapter 1). Helen, for example, referred to her planned rewatch of the first season of *Battlestar Galactica* with a friend as a marathon. Marathoning was also mentioned to catch up to a current season on DVD or online. Douglas, Buffy, Knitmeapony, and Elly, all quoted above, made reference to watching multiple episodes in one sitting. In talking about the first season of *The Wire*, Heresluck stated that once she was "officially hooked," she then purchased the DVDs "so that I could essentially sit down one weekend and mainline the back half of the season." Lauchita did the same in relation to an older HBO series *In Treatment*: "I downloaded all the episodes and watched them in two days." Helen, however, also engaged in this practice in relation to those series to which she felt *less* of a commitment. For example, she chose to marathon the past two seasons of *Top Chef* at the end of a semester. Anne C. made a similar point:

> Like right now my main fandom is *American Idol*. Everything else is just kind of noise to me. There are other shows that I enjoy watching but I am not fannish about it. So if I'm not fannish about it, I don't have the urgency and a lot of times I just prefer to mainline it and watch it from the beginning to the end. (Anne C.)

For others the displeasures of segmentation and interruption associated with the commercial broadcast outweighed the desire to see new episodes as soon as possible. But rather than using the live pause on the DVR or recording and playing back single episodes, these participants let a number accumulate and then marathoned them.

> I really enjoy not having commercials and I found that I prefer watching TV series in chunks rather than one episode a week. ... Typically I will collect a month's worth of stuff and then spend an evening watching 4 episodes of something. (Revan)

> Usually when it's a very suspenseful series and suspenseful content, one hour just doesn't cut it. So I would rather just watch 3 or 4 episodes in one go and then forget about it and then wait for another 3 or 4 episodes to be backlogged on my box. (Ellen)

While Revan and Ellen both used DVRs to break the paced weekly delivery of network television, Phillipe was a self-described "bulimic entertainment addict" who downloaded his content:

> It was terrible because I start one show/one episode and then okay, I'll watch the next, and the next one. So I can pass like 2–3–5 hours sometimes. A whole day watching my things. (Phillipe)

For series that were currently airing, he found himself caught between the push of anticipation and pull of the binge:

> When I watch them online, there is only one episode because the next one will be next week. … Sometimes I will restrain myself and for, I don't know—a month, I won't listen to some shows. So after a month I have like 3–4–5 episodes to watch back to back. (Phillipe)

To sum up, marathon viewers engage in the practice for reasons not limited to affective intensity.

Concluding Thoughts

Affective relations as I hope to have demonstrated, are central to patterns of viewing associated with Television 2.0. Committed viewing forms the basis of an affective relationship although, as we have seen, one can be a regular but not committed viewer of a series out of habit or because another household member has an emotional attachment to it. Picking up a new series used to be done in sync with the fall rollout by the networks, whether viewed live or time-shifted. While one could always catch up through repeat broadcasts or pick up an older series through syndication and then DVDs, the findings show that network sites, streaming subscription services, and VOD services provided by cable+ providers are popular means to both catch up and pick up. Once an affective relationship is established, intensity varies depending on a range of factors such as genre, the creators and actors involved in production, and bourgeois aesthetics.

The more intense the relationship and the more involved one is in fan communities, the more likely one is to anticipate new episodes and seek them out, using *whichever mode* provides the most immediate access. Live viewing of the original broadcast is the optimal means to this end. To view new episodes of current American or British series in continental Europe, South America, or Australia/New Zealand, however, or if one is a cord cutter in North America, one is generally required to engage in unauthorized downloading. Simply dismissing anticipatory fans as "pirates" frames downloading as a moral and ethical shortcoming rather than a structural limitation of the national broadcast model and a subsequent failure of the television export market. In addition, fans with more intense relationships are more likely to engage in repeat viewing of content, often to gain a more in-depth appreciation or understanding of a particular episode. Some are also collectors of DVDs regardless of whether they rewatch them or not. In the case of original series produced by Netflix and other streaming services such as Hulu and Amazon, marathon viewing becomes a form of anticipatory viewing.

It is important to keep in mind that not all fans are anticipatory or are anticipatory about all the series of which they are fans. Many are content to time-shift and playback a single episode to keep pace with delivery or to catch up at a later date by doing a marathon of episodes on their DVR, sometimes as a practice of resistance to commercial network practices. Others will be content to wait until the series is either available on DVD or, increasingly, as part of the digital archive provided by subscription streaming services at which point they may engage in some degree of "binge" viewing. Finally, as some of the data samples have illustrated in this chapter, fans often enjoy engaging in the range of these affective viewing practices with friends or family. In the final chapter, I will provide a detailed examination of a range of online participatory practices that are based on relations of affective intensity.

References

Bothun, D., & Lieberman, M. (2011). Speed of Life: Discovering behaviors and attitudes related to pirating content. PricewaterhouseCoopers LLP. Retrieved from http://www.pwc.com/us/en/industry/entertainment-media/publications.html

Bourdieu, P. (1993). *The field of cultural production: Essays on art and literature*. New York: Columbia University Press.

Bury, R. (2005). *Cyberspaces of their own: Female fandoms online*. New York: Peter Lang Publishing.

Evans, E., McDonald, P., Bae, J., Ray, S., & Santos, E. (2016). Universal ideals in local realities: Online viewing in South Korea, Brazil and India. *Convergence: The International Journal of Research into new media technologies*. Advance online publication. doi: 10.1177/1354856516641629

Fiske, J. (1989). Moments of television: Neither the text nor the audience. In E. Seiter, H. Borchers, G. Kreutzner, & E. M. Warth (Eds.), *Remote control: Television, audiences, and cultural power* (pp. 56–78). New York: Routledge.

Gray, J. (2003). New audiences, new textualities: Anti-fans and non-fans. *International Journal of Cultural Studies, 6*(1), 64–81.

Grossberg, L. (1992). Is there a fan in the house? The affective sensibility of fandom. In L. A. Lewis (Ed.), *The adoring audience: Fan culture and popular media* (pp. 581–590). London and New York: Routledge.

Jenkins, H. (1992). *Textual poachers: Television fans & participatory culture.* New York: Routledge.

Jenkins, H., Ford, S., & Green, J. (2013). *Spreadable Media: Creating value and meaning in a networked culture.* New York: New York University Press.

Jenner, M. (2016). Is this TVIV? On Netflix, TVIII and binge-watching. *New Media & Society, 18*(2), 257–273. doi: 10.1177/1461444814541523

Leaver, T. (2008). Watching *Battlestar Galactica* in Australia and the tyranny of digital distance. *Media International Australia, 126*(February), 145–154.

MacNeil, K. (2016). Torrenting *Game of Thrones*: So wrong and yet so right. *Convergence: The International Journal of Research into New Media Technologies.* Advance online publication. doi: 10.1177/1354856516640713

Mittell, J. (2006). Narrative complexity in contemporary American television. *The Velvet Light Trap, 58*(Fall), 29–40.

Steirer, G. (2014). The personal media collection in the era of connected viewing. In J. Holt & K. Sanson (Eds.), *Connected viewing: Selling, streaming, & sharing media in the digital era* (pp. 79–95). New York: Routledge.

Vonderau, P. (2014). Beyond piracy: Understanding digital markets. In J. Holt & K. Sanson (Eds.), *Connected viewing: Selling, streaming, & sharing in the digital era* (pp. 100–123). New York: Routledge.

· 5 ·
FANDOM 2.0
Six Degrees of Participation

The attentive and anticipatory viewing practices discussed in the last chapter serve as the springboard for all participatory fan practices. Participation and participatory culture, the focus of this final chapter, have been studied extensively since the publication of *Textual Poachers* (Jenkins, 1992), and television fans and fandoms continue to form the backbone of fan studies.[1] Yet fan studies has never paid attention to those fans "who merely love a show, watch it religiously, talk about it, and yet engage in … no other activities" (Gray, Sandvoss, & Harrington, 2007, pp. 3–4). Part of the reason for this lack of attention is the persistence of the notion that community is at the heart of fandom. Jenkins, for example, is primarily concerned with demonstrating the ways in which collective engagement has contributed to the formation of specific fan communities as well as a larger participatory culture. The effect, I argue, even if not the intention, has been to operationalize a binary between participatory and "non-participatory" fans, measuring the latter against the former and finding them lacking (Bury, 2018b). While it is obvious that anyone who invests time into writing fan fiction or making fan videos (vids) has a strong affinity with a media text, those who do not cannot be assumed to have a lesser one. Reviewing the data samples in Chapter 4, I found no distinction in expressions of intensity between those active in fandom and those who were not. What's more, those involved in fan communities devoted to one series were not necessarily involved in the fandoms for other series. Elly, for example, compared *Hawaii Five-0* to *The Sentinel*:

> I really love the characters. I love the show. I think it's great. Even they have been in reruns right now. They have been rerunning a few and next week is going to be a new episode, I will watch the reruns. I don't want to miss the reruns either. But the fandom just holds *no* allure for me, whatsoever. My friend and I have been driving ourselves crazy trying to figure out. I mean, *The Sentinel* just *grabbed* me by the throat and shook me and there I am, I am in the fandom. I would be writing fan fiction for it too except right now I just don't have the time to do it. (Elly)

In order to accurately detail changing patterns of participation in relation to Television 2.0, this binary needs to be deconstructed. Abercombie and Longhurst (1998) developed an *audience continuum* to classify fans by activity and commitment to the text. They begin with "consumers" and then add three more categories: "cultists," "enthusiasts," and "petty producers." Rather than group fans into categories, an exercise that is difficult to support empirically, I make the case for a continuum of participation, with those practices that require lesser amounts of involvement in fandom on one end and those that are bound up with fandom on the other end. I begin this chapter with a short discussion of the survey data to provide a wide-angle shot of fan engagement in an online context. I then turn to the interview data to drill down into four clusters of practices: information seeking, reaction and collective interpretation, community making, and cultural production. I pay particular attention to the role of social media. As with streaming and downloading platforms, platforms such as Facebook and Twitter not only enable new ways to spread media content but serve to alter existing fan practices and create new ones in ways that further complicate the participatory continuum.

The Participatory Continuum

The survey asked the respondents how often they visited online sites, including websites, discussion forums, and social media platforms: forty-one percent did so frequently, twenty-six percent sometimes, twenty-two percent occasionally, and twelve percent never did so. When asked specifically about online discussion forums, thirty-seven percent reported having visited at least one, although less than half of those who did had participated in the discussions. As for social media platforms, forty-seven percent had "liked" a post about a series on Facebook (just under ten percent did not have a Facebook account), but only eleven percent had posted a comment on a Facebook page dedicated to a favorite series. Only one third of the respondents reported using Twitter for fan-related activities. The final cluster of questions concerned cultural production. One quarter had read fan fiction and thirteen percent had written it and shared with others online. The gap between "consumption" and "production" of vids is even more striking: whereas thirty-one percent had viewed at least one vid, only three percent had produced at least one and shared with others online (six percent were unfamiliar with either type of creative work). As for statistically significant demographic differences, Twitter was used more by women, the eighteen to twenty-nine cohort, and by those respondents residing in the United States. Women and those living in the United States and United Kingdom also read and wrote more fan fiction.[2] Taken together, these findings suggest that the majority of fans of popular television can indeed be classified as participatory but that most are clustered on the "less involved" end of the continuum, doing more than viewing but not directly engaging in the hallmark practices associated with community and participatory culture.

Information Seeking

Fiske (1987) argues that television texts need to be understood in terms of primary and secondary relations: the pleasures of viewing the primary text are extended by a range of secondary texts about the series, its actors/stars, writers, and producers. Jenkins gave the example of fans in the pre-internet days scrutinizing previews that aired at the end of new episodes and "[racing] to buy TV Guide as soon as it hits the newsstands" (1992, p. 57). Bennett (2014) reiterates Jenkins' claim of the importance of shared knowledge to fans and emphasizes the affordances of the internet and social media in spreading such information. All 72 participants were involved in information seeking.[3] Some were looking for "Wikipedia type generalities about the series" as was Nomdeplume for *The Wire*. Others like Annika "will just get curious about one of the characters or the setting or any sort of detail like that and decide to go look it up." Others wanted to identify the actors who appear familiar from other roles:

> I wanted to find out who was playing Drogo [*Game of Thrones*] because he was so enormous and I was like, who the hell is that guy? So I tracked him. He is a Hawaiian actor [Jason Momoa] who has been in *Stargate Atlantis*. I know him; he's the big dreaded guy. ... It's him! (Khal)

Other kinds of information being sought included spoilers, previews for upcoming episodes, and the "extra features" that used to be exclusively found on DVDs. A number of participants mentioned network sites as their starting point for a search. Others went to specific third-party news sites. LWR, for example, used "a Sci-Fi news service called Blastr. ... I have an iPad and iPhone app that I check that has news about Sci-Fi and I also go to their website from time to time." Douglas' commitment to "horrible, cheesy reality shows" motivated his information seeking: "I love *The Jersey Shore*; I love *Big Brother*; I watch *Survivor* still even though I love it less; I love the *Amazing Race*. So I follow a lot of that kind of stuff online." Several others, including Annika and Liz, mentioned doing information searches using Google or sites like IMDb while viewing an episode of a favorite series. Liz, for example, said that it was useful to get background on a character she was not familiar with. She also sought information to catch up if she began viewing partway through a broadcast. Will gave a different reason for seeking information: to keep up with a series (*The Event*) in which he was losing interest but did not want to give up on completely.

A number of participants expressed interest in seeking out and reading reviews and recaps, a demonstration of their investment in bourgeois aesthetics (see Chapter 4):

> There is a site called HitFlix which has some television critics that work for them. One fellow's name is [Daniel] Fienberg and another fellow's name is Alan Sepinwall. They have kind of got a little community and they all know each other across the country and they will all blog about different shows that they like critically. So I like kind of reading the different perspectives. There are probably 5 or 6 that I will read. (Camden)

"Liking" on Facebook and "following" on Twitter provide new means of acquiring information about a favorite series. The customized newsfeeds are "push oriented," serving to transform "pull-oriented" browser-based information-seeking practices into information receiving and aggregation (Bury, 2018a):

> I like a lot of shows. It's so funny because I feel like my newsfeed, like when you sign in it's probably like fifty bordering on sixty percent updates from all the different shows that I've liked. I like that because I get up-to-date information on what's happening with these shows. (Tasha)

> I get most of the information that I need just passes through my Twitter feed and then I get to follow those links to wherever they go. For instance, I follow Jason Mittell [TV scholar] and he will have a link to something like that on there and then I follow the link through and then I am at Entertainment Weekly reading their story about it. (Douglas)

The appeal of these platforms is not only the convenience but the immediacy of information afforded by the customized newsfeed:

> Twitter and Facebook are nice for immediate information about the stars themselves and I appreciate the TV stars. (Mary)

> [Twitter] is more instantaneous. ... I have got news sites on there but I have also got fandom people on there and you see a lot of useful links and news come by. (Vera)

> There are a couple of news ones that I follow that have *every* single thing that Adam Lambert or Kris Allen [*American Idol*] does. They will have a link to it or tweet about it. So I follow, like I get a lot of fandom information that way. (Anne C.)

It is worth noting that these emerging practices related to information aggregation make visible to one's social networks the affective relationship that one has to particular texts. Such relationships remain hidden when seeking out information on a website or "lurking" on a discussion forum or the comment section of a blog or recap site.

Social media platforms also blur the line between information and interactivity: instead of going to a website and then to another site or platform, such as personal email, listservs, and discussion boards, one can easily share and discuss information with one's Facebook friends or Twitter followers:

> We @ at each other and links things to each other: did you see this? That kind of thing. Part of it is of being able to have live conversation about things that come out. New information comes out and you are able to talk about it. I get my information from Twitter and then I am able to retweet and point other people to it and talk about it in that space rather than sort of like going to email or something like that. (Karen)

The majority of participants who used Twitter, however, felt that the 140-character limit made it unsuitable for interaction. As Diva stated, "I find it a very useful information service. But I don't understand how people can use it to communicate with each other."

Facebook and Twitter also provide a "straight from the horse's mouth" perspective not associated with print or the web. "Public figures can now seemingly speak directly to their fan base without news or management filters" argues Lucy Bennett (2014, p. 8). Several participants spoke enthusiastically about following the feeds of TV actors and producers:

> If Nathan Fillion is tweeting from the site of *Castle* then I am like, "ahhh what's he's doing and this is so cool!" The guy who played Rodney McKay on *Stargate* [*Atlantis*] is also a big Twitterer and he is hilarious. And Stephen Fry of course. I mean, I am a *Jeeves and Wooster* fan all the way up to him being on *Bones*. Just a HUGE fan of that man. He is *massive* on Twitter. (Knitmeapony)

> I follow Brent Spiner on Twitter and he tends to post a ton of things where fans ask questions about what he is doing, where he is, his experiences of playing Data [*Star Trek: The Next Generation*]. He is very sarcastic and very witty which is interesting. (Jayne)

> I follow—oh what' his name? Sutter, Kurt Sutter. He's the guy that created and writes *Sons of Anarchy*. So he tweets a lot about the show and what's happening and where they are up to and so on with making the show. He also blogs about it. He follows some of the main players so I find that really interesting to get these kind of insights into what's going on as they produce the show. (Khal)

The possibility of communicating directly with showrunners, actors, and celebrities was an exciting prospect for participants such as Anne C., who saw it as "the fourth wall ... breaking down." Yet the TV 2.0 data revealed no evidence of such a large-scale shift. Of the few participants who mentioned tweeting an actor or showrunner, there was no sense that a reply was received or even expected:

> Joe Mantegna [*Criminal Minds*] said something about one of his movies being out and he simply tweeted, it will be out on such and such a day and I tweeted back, "I have already pre-ordered mine." Yeah that kind of thing. There was something about supporting veterans and I said, "Thank you. My father's a veteran." (Mary)

Mary had the following to say to fans who expected replies: "Get it over it people. ... I don't expect somebody who is starring in a television show to send me a personal reply to anything. That's unrealistic. I appreciate what they do." Libby had received a reply from showrunners Hart Hanson (*Bones*) and Shawn Ryan (*Chicago Code*) when she asked whether viewing on TV or online affected the amount that producers and actors were paid: "I admit, it was pretty exciting when they actually wrote back to me. ... I got all geeked about it and I was a little embarrassed." Buffy tweeted a producer of *So You Think You Can Dance* for a different reason:

> I love dance and I obviously love a show where I get to watch it for two hours a week. But there are a lot of things about it that infuriate me. So sometimes Twitter can be space where I will go, like Nigel Lythgoe the producer is on Twitter so every time that I am displeased with something I feel like I am really taking control when I go and tweet him about my displeasure!! (Buffy)

Marwick and Boyd (2011) suggest that fans do not tweet celebrities to receive a reply as much as to display an affective relationship, or in Buffy's case to voice a criticism. Thus social media, Twitter in particular, serves to alter established mediated producer–fan and actor/celebrity–fan relations and provides a new set of fannish pleasures even if they do not necessarily involve direct communication (Bury, 2018a).

What Do You Think?

In addition to secondary texts, Fiske (1989) contends that the primary text is surrounded by *oral culture*. Talk about television is an integral part of the activation of meanings and pleasures of the polysemic text. It can be an important component of viewing among members of the household and/or friends in the living room.

> I know it sounds stupid but I enjoy watching [reality shows] live and talking through them. Like, I talk to my partner when we are watching it. (Farah)

> Even though TV is more of an individual thing I like to share it with somebody. I like to watch it with somebody else. I don't talk through the whole movie; I don't do any of that. But just to look at somebody after and share that moment where they hated it or they loved it and discussion and everything. I find every time I watch something by myself I end up going, "I wished I'd watched that and been able to share that moment with somebody else." (Douglas)

For Ellen, however, talk detracted from the meaning-making process. Although she enjoyed watching reality series and events such as *The Oscars* with others as a social experience, she added, "usually I feel like I need to focus and I need to keep paying attention to the storyline. I kind of want to get lost in that show as well." The next sample illustrates the role of the second screen in accommodating different interactional preferences while co-viewing:

> There really is nothing funnier than fifteen Whovians [*Doctor Who* fans] gathering in front [of the TV] and everyone is on their own laptop *while* we are watching the show. I have sat there and like and having an IM conversation with someone who is sitting on the other end of the couch because we enforce no talking rules sometimes during some of these shows. (Knitmeapony)

As I pointed out in the Introduction, the emergence of ICTs afforded the formation of a vibrant online fan culture to complement, and indeed extend, the domestic oral culture of television. Only a few of the participants were old enough to have interacted with other fans in the first fan forums.

> When I got to college in the late 1980s, I got on to Usenet. That's when things really blossomed for me, on a lot of levels. I came to social maturity on Usenet and the local BBS system at my college ... Had great deep and/or silly conversations with people about *Star Trek, Doctor Who, Monty Python*—and later, *Red Dwarf, Buffy*, and on and on. (Willow)

A few participants mentioned listservs (e.g., Elly and Mary) but most made reference to the "next generation" of message boards and websites, administered either by fans or by the broadcasting networks, including Yahoo! Groups, fanforum.com, EZ Boards, and Television without Pity.

> I have been reading recaps and going to forums on Television Without Pity for 8–9 years. That's sort of my primary forum site. ... Also I tend to go to the [TWOP] forums for whatever sort of shows I am currently watching that are currently airing and sort of checking out real time responses and things like that. (Margene)

> It was 2000 when I actually got access to the internet properly and then I joined the BBC Cult message boards which had an *X-Files* forum on it. And that was when I started writing more fan fiction, talking to other fans and discussing episodes and stories and stuff like that. (Nem)

> I think I was looking for a new [*Dollhouse*] message board to post on. I joined it and was active enough there that when they were looking for people to be involved in moderation and helping administer it, they sort of invited me to do so. So that was fairly standard in terms of a message board. There was episode discussion and spoiler discussion and there were people on there who creating fan art and banners, etc. (William)

These samples provide a range of engagement from "read-only" (Margene), reading and posting (Willow, Nem), to running a board as William did for a period of time.[4] For those who engaged in the first practice, reading the comments of others was a means to validate meanings already made and to learn about new details and perspectives. Phillipe said that when he watched a new series he went online "to see what people said about it. Is it good? I actually watched the first episode and then I went online to see if my reaction ... I like to check my reaction." Suzie noted that going on a message board for *Lost* helped her pick up on things that she had missed while viewing. Courtney expanded on the value of the opinions of other fans:

> There was one [*L Word*] blogger that had a synopsis every week or every episode and it was outstanding with pictures and so on and commentary. And the boards for the opinion and the commentary and the contemporary issues, that was a really, really vibrant community. I think I tried to get a log in and maybe post in a couple of cases but mainly I was just reading all the other commentary. It provided a whole separate dimension to the whole [viewing] experience. (Courtney)

Social media has extended opportunities for reaction and interpretation, both among people one already knows and those one does not. Interestingly, very few participants mentioned using Facebook for this purpose.

> Everyone [was] taking to Facebook with their outrage at how bad that last episode was or how disappointed they were in where it went or happy, whatever the case may be. But I remember my feed turning into this string of responses to *Lost*. (Buffy)

According to Bennett (2014), one reason for not discussing episodes on Facebook is a concern for spoilers. Buffy said that when she wanted to find out what her friends thought of an episode, she would post a comment, "trying to make it as oblique as possible without revealing details … and then hoping your friends will respond in turn." Twitter on the other hand was valued for providing the possibility for immediate reaction at the time of viewing, a practice enabled through use of a second screen. According to the Nielsen Company (2016c), in 2015, almost one billion tweets were sent about television in the United States. A survey by the Pew Research Center found that twenty percent of those who used their phones as a second screen reported that they checked to see what others were saying about the program that they were watching and nineteen percent posted their own comments about the program (Smith, 2012). Fourteen (one fifth) of the participants followed television-related hashtags during a live broadcast. Will noted that he used the hashtag for a current affair program that aired on the BBC on Thursday evening: "I find lots of my colleagues in those tweets. So it's a social thing as well." Camden also enjoyed looking up the hashtags that were promoted on Jimmy Fallon.

The next set of data samples capture the pleasures of what Page (2012) refers to as *ambient affiliation*, and as such signals a connection to a larger group that does not necessarily involve interactivity:

> I know there was quite a few a few weeks ago when *Supernatural* did their Super Uber Meta episode, as they billed it. So there was a ton of live tweeting of that one because everyone was just so excited and wanted to, I don't know—disseminate as much as possible and as quickly as possible and that seemed to be the way to do it. I don't remember the hashtag they used but it was trending pretty quickly and pretty high. (Rene)

> If you follow a hashtag on Twitter, like there is, I mean, I have been part of global "oh, come on's"!! Where literally there are people and I have no idea who they are. They live in Montana, that's awesome. And we are collectively going, "Tony DiNozzo [*NCIS*] would never do that!!!". … I am not sure that I would have a satisfying TV experience if I couldn't immediately keyboard smash at people and go, "Oh, God! How is this going to work?!" Or, "That was brilliant!!" or "That was terrible!" (Knitmeapony)

Page's study results suggest that what appears to be conversational may in fact be "para-social simulations of conversationality found in broadcast talk" (p. 184)—in other words, people talking at each other rather than to each other. Similarly, Yvette Wohn and Eun-Kyung Na (2011) analyzed over 1000 tweets made during an October 2009 broadcast of *So You Think You Can Dance*. They mapped tweet patterns onto the content of the show, noting spikes in emotion, attention, and opinion messages during commercial breaks, and when one of the dancers was injured. The rest of the program was dominated by informational tweets. They found little evidence of interactivity: less than four percent of the tweets directed at a specific user using the replay, i.e., @ symbol, were reciprocated. The only TV 2.0 participant who claimed to interact during a broadcast with other fans who he did not know was Khal:

> [My wife and I] use Twitter quite a lot while we watch the show. So we make comments and so on when we are watching the show. It turns out there are other people around the country that watch television the same way we do. So for instance, when *Top Chef* comes available … you can have a conversation about it as you watch it. (Khal)

Jenkins argues that online forums provide the opportunity for viewers to become active in participatory culture: "more and more of them are sneaking a peak at what they are saying about the show on Television Without Pity, and once you are there, why not post a few comments. It's a slippery slope from there" (2007, p. 361). Today this idea can be expanded to include social media: sneaking a peak at a hashtag and then sending out a few tweets. Analysis of the TV 2.0 data, however, suggests that a "line in the sand" metaphor is more appropriate. While some of the "read-only" participants stated that they felt uncomfortable or shy about posting or that they had nothing further to contribute, the main reason is captured by the next data sample:

> I will [talk about TV] face-to-face: at work, at school with friends but it's not the thing I will do online. … I may sometimes if I am amazed or if I am really touched by something, communicate it on Facebook and then if some people respond I will respond to them. But that's as far as I will go. (Phillipe)

Watching and talking about the show with family and friends, in combination with the read-only practices of information seeking and validating one's reactions and interpretation, was thus considered to be sufficient. Even William described his fandom as "personal" rather than participatory despite his short-lived foray into forum administration. Conversely, not having friends or family with whom to talk about a particular gender or series was a driver for a few participants to seek out other fans online. Tasha noted that she did seek out community online but only for shows that she did not watch and talk about with her husband or friends. Suzie similarly described having "an urge to communicate with people when I am watching television":

> My husband and son they watch *Amazing Race* and *Survivor* but they don't watch *Real Housewives* and when I watch *The Bachelor* they say, "what is that!" So I had nobody here to talk to about these shows so I think it's really nice that you can find a community of people who care about the show and talk about the show and they are just at your fingertips. (Suzie)

I concur with Jenkins that the increased visibility and accessibility afforded by ICTs and social media does provide multiple entry points into fandom. However, the TV 2.0 survey and interview findings suggest that many prefer to take pleasure in ambient affiliation without becoming an active member of a fan community.

Community Making On (the) Line

Before the internet existed, Bacon-Smith (1992) made the case that fan communities could be divided into "interest groups" and "circles." The first were composed of up to 500 members (an estimate that is not backed up with empirical evidence). While some may have interacted via mail and/or face to face, many only knew of each other through "reputation" in the fandom (e.g., as fan fiction writers). Such groups provided a sense of unity but were "far too large to meet the personal needs of its members for close connection" (p. 26). These needs were met by much smaller groups, which relied heavily on regular face-to-face communication. Asynchronous computer-mediated communication blurred this boundary, creating the conditions for the formation of interactive communities beyond the size limitations imposed by corporeal presence in a common physical space as well as at a common time. Drawing on Judith Butler (1990) and her notion of performativity, I argue that what gives any community its *substance* of legibility is the routinized, regularized repetition of a variety of acts or practices (Bury, 2005). In text-based cyberspaces, these acts are based on a hybrid form of communication—speech that has been "frozen into artifact" (Turkle, 1995, p. 183). Just under half ($n = 33$) of the interview participants identified as being or having been members of at least one specific group in which individuals have direct contact and the possibility of regular interaction with each other:

> Basically, I remember, actually a lot of the fandomy discussion was through Yahoo! groups. I would join these and discuss that way and make friends, etc. We would use the Yahoo! groups to really *talk* about fandom ... and participate in that way. (Karen)

In addition to the topic-based discussions on series and fan creative works, a number of participants also spoke of personal connections:

> I did have [a sense of community] in JAG fandom because I was on a mailing list and you saw more personal things from people and you could discuss things. (Vera)

> I think that back in the Usenet day it really was community. It was people discovering each other, really fairly deeply intertwined in each other's lives. (Revan)

Elsewhere I have made the case that friendship is integral to online fan community and, in some cases, outlasts interest in the primary text around which the community formed (Bury, 2008). As Mary said, "the longer the group has been around, the more there is a sense of family and community." This sense of connection and friendship was carried over from the older listservs and message boards to LiveJournal. This early social networking service was launched in 1999 and adopted on a wide scale around 2003 specifically by those fans who produced their own creative works (Coppa, 2006). For a number of participants who migrated from the older platforms, LiveJournal became and remains their fannish home:

> So we might not actually make a comment on each other's journal for 6 months but then we do. And we still have this continued history that is probably like 6 or 7 years. It's the people who are doing the same kind of thing that I am doing who I probably feel close to than the people who were just that quick fandom person. (George F.)

> A bunch of people that I have never, ever met and I probably never will. But yeah, you kind of get to know everybody or at least who they are sort of within the community. So yeah, it's nice. (Tarsus)

I found little evidence of community making on the social media platforms however. As I pointed out in the previous section, the character limit baked into Twitter's design was a contributing factor. Those active in fan communities such as Diva and Heresluck pointed to its unsuitability for in-depth, sustained interaction, contrasting it to LiveJournal:

> I am much more interested in seeing someone's post about it a day or two later when they've had a chance to organize their thoughts and produce something sort of interesting and in-depth. That's my priority. I think this partly has to do with my having a very small circle of people on Dreamwidth and LiveJournal who I read. ... I would

rather wait until someone has time to do something a little bit more long form than see whatever random 140-character thing they posted to Twitter. (Heresluck)

Karen still went to LiveJournal for "fan fiction and other sort of community type activities." As for Freda, "fandom happens on LiveJournal … or increasingly for me, on Twitter." When asked how she found fans to follow she replied, "Most of the people I follow on Twitter I already met on LiveJournal first." Of course work-arounds are always possible with the use of the @ function as already noted. Anne C. summed it up best: "You can build a community though if you are willing to follow a lot of people. I don't use it this way."

Although Facebook's architecture is similar enough to that of LiveJournal to make it seem suitable for community making, this has not turned out to be the case:

> My Facebook is more like my personal life and my family and friends and I kind of don't intersect it with my fannish life. (Margaret)

> Facebook is where I keep up with people that I know from college and graduate school and with certain people who are current colleagues and were former students. So who I met in an offline capacity, although because some of the people that I met through fandom are now personal friends, I am also connected with some of them on Facebook. (Heresluck)

Facebook is premised on having as large and wide-ranging a social network as possible, hence the algorithm-generated list of "people you may know" with the weakest of ties (friends of friends) to whom one is encouraged to add to one's network. Although it is technically possible to set up separate friend lists, none of the participants mentioned doing this. That said, given the importance of friendship in fan communities, Facebook provides an opportunity to supplement and extend those connections:

> Because some of the people that I met through fandom are now personal friends, I am also connected with some of them on Facebook. (Heresluck)

> To be Facebook friends with someone who you are in fandom with, to have the privilege of having their real name and contact information which can happen and that does happen and people bond and stuff like that. It does add more depth to it. (Stevie)

Facebook accounts devoted to a particular fandom were mentioned by a couple of the participants in relation to regular communication and connection:

> I had first of all joined a group on Facebook which was the London *I Want to Believe* get together. It was a big group of fans who would write, "If there was a London premiere of

The X-Files [movie] we are going to be there." So I was like, right, well that sounds good to me. So I joined that and then joined various other X-File groups on Facebook. (Nem)

Everybody just gets to know everybody else. There will be times when people will just have little jokes with each other. You get to kind of know everybody's preferences and you will kind of go, "Wow! Denise, look at this photo." It's usually Zachary Quinto [*Star Trek: Into Darkness*] just being fabulous! "Look at this. Isn't he wonderful?" And she will go, "Yes, isn't he wonderful." And then everybody will kind of join in. It's just really knowing what everybody kind of enjoys and everybody is going to find funny. (Tarsus)

In Nem's case, the ensuing friendships she developed with other group members did not come from Facebook interaction alone but from the meet-up at the screening of *The X-Files* film. Interestingly, she found out about LiveJournal through Facebook and it is the former that she says has become the center of her fandom. Tarsus found about the *Star Trek* reboot Facebook group through a Yahoo group, but it is not clear if she already knew some of the members when she joined. To sum up, ICTs both enable and disable community making among participatory fans (Bury, 2017). While the older platforms made extended, sustained interaction possible on an unprecedented scale, social media platforms have not replaced them, in part, because they do not lend themselves to such interaction.

The Producers

Although all of the interview participants who had community involvement had read fan fiction or viewed vids, only six identified as writers—Anna B., Elly, Mary, Notesofwhimsey, Robert, and Vera—and three as vidders—Anne C., Heresluck, and Stevie. In the early days of the internet, Elly recalled having to use one mailing list for posting fiction and another to comment on it. LiveJournal has made the sharing of creative fanworks much more convenient:

You would post your fanfic to the community or post a link to it on the community and that's how people would find out. Or trust that your followers would tell other people about it or rec [recommend] it or mention it in their posts or something like that which is actually more effective than it sounds. (Stevie)

I post an announcement to my own Dreamwidth account which mirrors to LiveJournal. I post an announcement in the vidding community on Dreamwidth and LiveJournal. Then, depending on what the source is I may also post to other places. So there are sometimes fandom-specific vid communities. (Heresluck)

People will announce them on LiveJournal but it will be embedded from another video service. (Vera)

Revan described the creative communities that formed on LiveJournal as "really dynamic": "So everybody is just kind of in the same space together reading and commenting on each other's stuff." The circulation of creative works through the web and the journaling platform is part of a gift economy—readers and viewers are not expected to reciprocate in kind: "what the gift in the digital age requires for 'membership' into the fan community" argues Paul Booth, "is merely the obligation to reply" (2010, p. 134). This reply could take the form of a positive comment or a few lines of constructive criticism (Jenkins, Ford, & Green, 2013).

That said, finding out about and becoming a member of these creative LiveJournal communities was not straightforward: Vera joined when an invitation was required: "It took a while but eventually I got invited to a rec com. ... Somebody started friending me and then you start friending more people and you get this community and I like that a lot."[5] Zee didn't need an invitation when she joined in 2004 but she had been "hanging out on the shipping boards."[6] Anne made a similar point about gaining access to the vidding communities:

> My roommate had been in fandom for years and when I got that obsessed about it, she actually had some fanvids for *Queer as Folk*. She showed me some of those and then I went crazy trying to find as many fanvids as I could. As I spent a lot of time watching fanvids, I was in the fandom for a good couple of months before I, like maybe about 5 months before I even found fanfic. (Anne C.)

These samples confirm Busse and Hellekson's contention that LiveJournal has created its own set of exclusions: "LJ's software makes it easy to invite people in, but it also makes it easy to shut them out" (2006, p. 15).

Content sharing platforms such as YouTube and Vimeo, however, have altered the gift economy, enabling what I refer to as a second order of spreadability (Bury, 2018a). As a result, "fan producers are no more able to control the dissemination of their texts than corporate producers" (Russo, 2009, p. 127). Freda, who did research on vidding as a doctoral candidate, noted "a tension in the community [as to] whether vidding can exist outside of LiveJournal." She admitted to being "a bit of a snob": "A lot of times if I see a vid and I don't know the vidder or, and this sounds really embarrassing, or if the vid is on YouTube I won't watch it." In contrast, Anne C. was comfortable posting her vids to YouTube, although she also noted that "there's also sometimes some snobbery involved where it's like if you post a YouTube you are not as good." Yet this second-order spreadability also can extend the pleasure of the primary text for fans who are not members of the creative communities:

> I think the first [vid] I ever saw was for *Lost*. I think my roommate actually showed it to me. It was actually a funny one. They cut together—one of the characters, Hurley, said "do" like all the time. Someone went and put together every single time he had said "do" in the first season. It was pretty funny. That's the first one I remember seeing. I didn't even realize that folks did that until I saw that and I was like, that was pretty good. (Anna B.)

> Every now and then a friend of mine will post something on Facebook, for example, a fan vid but they don't know it's a fan vid. They don't know what it is; they don't know about fandom. It's like, "I found this cool video. It's got House in it" or whatever is going on that day. At which point I usually have to educate them as to what it is they've found. "Oh, look, you found a fan vid. This is what a fan vid does." (Rene)

Rene also chose to post the occasional fan vid to Facebook "if I think it's a really good one."

Gene, a communications professor, knew little about participatory fandom until he was asked to review a scholarly article which analyzed fan-made reedits using scenes of a gay couple from *Verbotene Liebe* (Forbidden Love), a German soap with English subtitles. When he found himself on extended bed rest, he went back to YouTube and watched all the original episodes of the soap and actively searched for other fanworks not only featuring the couple, but other queer storylines from international and American series:

> So it's been really exciting for me because it's the experience of being a fan, which I've never been or really understood. I mean, my earlier partner was a soap opera watcher and I would leave the room. I just couldn't stand the sound of it. (Gene)

Not everyone, however, had the same appreciation for fan creativity:

> There are a lot of [vids] on YouTube. To be honest with you, as far as things go for me. YouTube is pretty much disposable. You know, you watch it and you forget you watched it. So I might see a fan video, a *Doctor Who* fan video or this fan video or that fan video and oh, that's cute and laugh, laugh, and then forget I watched it. (GSM)

Thus the ties that have bound cultural production to fan communities are being loosened by the same Web 2.0 mechanisms that afford IPTV and the spread of primary texts.

Concluding Thoughts

As I hoped to have demonstrated, many fans with an affective relationship to a primary television text do more than just view the series regularly. The TV 2.0 findings, however, support my argument that previous research, including my own, has conflated participation with participatory culture, and overrepresented the fan positioned at the "most involved" end of the continuum. The internet and social media have without question made fandom more visible and accessible in terms of discussion sites and creative works. However, it would appear that any uptake in participation is most likely to take place at the "least involved" end. Information seeking using the web and information aggregation through liking and following is clearly the most popular participatory practice. Many fans also enjoy reading recaps and reviews as well as the comments on discussion forums, some posting on occasion, often with the intention of "checking" their own reactions and interpretations against those of others. These fans are not likely to take the next step and interact on a regular basis with other fans they do not already know and become members of online communities for the reason that they feel that their interactions with people they already know—family, friends, and colleagues—are sufficient. For those fans who do wish to build connections with the larger fandom, computer-mediated communication afforded by ICTs has enabled online communities to form beyond the size limits of local clubs or "circles" by creating the possibility of direct interaction within larger interest groups. Platforms such as listservs, news groups, discussion boards, and LiveJournal all enable sustained, regular interaction required for community making.

Social media platforms are without question having an impact on fan practices along the length of the continuum. First by liking and following, affective relations to television texts are made visible. Second Twitter in particular is blurring the line between viewing, reaction, and interactivity with the use of hashtags and a second screen. By following a hashtag, one can have a sense of ambient affiliation with other viewers one does not know but not necessarily a sense of community. Because of Twitter, fans are also able to receive direct information from writers, actors, and reality stars. Although one-on-one communication is possible, the practice remains one-sided: fans display an affective relationship by replying to tweets even if they are unlikely to receive replies or get followed in return. When it comes to community making and interactivity, social media platforms *disable* such practices due to their architectures and functionalities. Facebook and to some degree Twitter, however, need to be recognized for their role in supporting the maintenance and extension of already established community relations and friendships forged through fandom. Finally downloading and streaming platforms enable the spread of not only primary television texts but fan-created secondary texts as well. Thus, they effectively breach the boundaries of the creative communities and disrupt the established gift economy of sharing fan fiction and vids among members of those communities.

Notes

1. A sample of the monographs and edited collections influenced by early fan studies scholarship include Baym (2000), Booth (2010), Bury (2005), Gray, Sandvoss, and Harrington (2007), Hellekson and Busse (2006), and Hills (2002).
2. I did not include a question on fan activism, an unfortunate omission that reinforces the privileging of cultural production over activism (Bennett, 2014).
3. To ensure that the voices of the 39 "non community" members were adequately represented in this chapter, I have only presented data samples from them in relation to information seeking.
4. I do not use the term "lurker" due to its negative connotation.
5. "rec com" is an abbreviation of recommendation community. On LiveJournal, this would be a community journal for a particular fandom where members share their recommendations for fiction and/or vids.
6. "Shipping" is a term used by fans to describe the romantic pairing of two characters who do not have such a relationship in the series canon, that is, as written by the producers.

References

Abercombie, N., & Longhurst, B. (1998). *Audiences: A sociological theory of performance and imagination.* London: Sage Publications.

Bacon-Smith, C. (1992). *Enterprising women: Television fandom and the creation of popular myth.* Philadelphia, PA: University of Pennsylvania Press.

Baym, N. K. (2000). *Tune in, log on: Soaps, fandom, and online community.* Thousand Oaks, CA: Sage Publications.

Bennett, L. (2014). Tracing textual poachers: Reflections on the development of fan studies and digital fandom. *Journal of Fandom Studies, 2*(1), 5–20. doi: 10.1386/jfs.2.1.5_1

Booth, P. (2010). *Digital fandom: New media studies.* New York: Peter Lang Publishing.

Bury, R. (2005). *Cyberspaces of their own: Female fandoms online.* New York: Peter Lang Publishing.

Bury, R. (2008). Remotely embodied friendships in female fan communities. In S. Holland (Ed.), *Remote relationships in a small world* (pp. 174–198). New York: Peter Lang Publishing.

Bury, R. (2017). Technology, fandom and community in the second media age. *Convergence: The International Journal of Research into New Media Technologies*, 23(6), 627–642. doi: 10.1177/1354856516648084

Bury, R. (2018a). Television viewing and fan practice in an era of multiple screens. In J. Burgess, T. Poell & A. Marwick (Eds.), *Sage handbook of social media* (pp. 372–389). Thousand Oaks, CA: Sage Publications.

Bury, R. (2018b). "We're not there." Fans, fan studies and the participatory continuum. In M. A. Click & S. Scott (Eds.), *The Routledge companion to media fandom* (pp. 123–131). New York: Routledge.

Busse, K., & Hellekson, K. (2006). Introduction. In K. Hellekson & K. Busse (Eds.), *Fan fiction and fan communities in the age of the Internet: New essays* (pp. 5–32). Jefferson, NC: McFarland & Co.

Butler, J. (1990). *Gender trouble: Feminism and the subversion of identity*. New York: Routledge.

Coppa, F. (2006). A brief history of media fandom. In K. Hellekson & K. Busse (Eds.), *Fan fiction and fan communities in the age of the Internet* (pp. 41–59). Jefferson, NC: McFarland & Co.

Fiske, J. (1987). *Television culture*. New York: Methuen & Co.

Fiske, J. (1989). Moments of television: Neither the text nor the audience. In E. Seiter, H. Borchers, G. Kreutzner, & E. M. Warth (Eds.), *Remote control: Television, audiences, and cultural power* (pp. 56–78). New York: Routledge.

Gray, J., Sandvoss, C., & Harrington, C. L. (Eds.). (2007). *Fandom: Identities and communities in a mediated world*. New York: New York University Press.

Hellekson, K., & Busse, K. (Eds.). (2006). *Fan fiction and fan communities in the age of the Internet: New essays*. Jefferson, NC: McFarland & Co.

Hills, M. (2002). *Fan cultures*. London; New York: Routledge.

Jenkins, H. (1992). *Textual poachers: Television fans & participatory culture*. New York: Routledge.

Jenkins, H. (2007). Afterword: The future of fandom. In J. Gray, C. Sandvoss, & C. L. Harrington (Eds.), *Fandom: Identities and communities in a mediated world* (pp. 357–364). New York: New York University Press.

Jenkins, H., Ford, S., & Green, J. (2013). *Spreadable media: Creating value and meaning in a networked culture*. New York: New York University Press.

Marwick, A. E., & Boyd, d. (2011). To see and be seen: Celebrity practice on Twitter. *Convergence: The International Journal of Research into New Media Technologies*, 17(2), 139–158. doi: 10.1177/1354856510394539

Nielsen Company. (2016c). TV season 2015–2016 in review. Retrieved from http://www.nielsenadfocus.cn/us/en/insights/news/2016/tv-season-2015-2016-in-review-the-biggest-social-tv-moments.html

Page, R. (2012). The linguistics of self-branding and micro-celebrity in Twitter: The role of hashtags. *Discourse & Communication*, 6(2), 181–201. doi: 10.1177/1750481312437441

Russo, J. L. (2009). User-penetrated content: Fan video in the age of convergence. *Cinema Journal*, 48(4), 125–130. doi: 10.1353/cj.0.0147

Smith, A. (2012). The 'rise' of the connected viewer. Pew Research Center. Retrieved from http://www.pewinternet.org/2012/07/17/the-rise-of-the-connected-viewer/

Turkle, S. (1995). *Life on the screen: Identity in the age of the internet*. New York: Simon & Schuster.

Wohn, Y., & Na, E.-K. (2011). Tweeting about TV: Sharing television viewing experiences via social media message streams. *First Monday, 16*(3). Retrieved from http://journals.uic.edu/ojs/index.php/fm/article/view/3368/2779

CONCLUSION
Rhizomatic for the People

In the Introduction I stated that Television 2.0 was not a medium in transition but a hybrid, rhizomatic assemblage that dates back to the days of radio. Like all assemblages, it has a core that has formed over time, in this case a broadcast model of centralized transmission and privatized transmission. More accurately, larger economic, national, institutional, and corporate pressures not only created but maintain a substance of stability and coherence. As a consequence the markers and practices of what we associate with broadcast television are very much in evidence. Production is still concentrated in the hands of a few producers, controlled by either large media conglomerates or state apparatuses. Public and private network infrastructures still broadcast most of the content produced, organized around a twenty-four-hour programming day and paced delivery of the serials and series. Most programming is still transmitted over the air, or via ASL/DSL cable and satellite even if the number of paid subscriptions for such services is dropping gradually in the Canadian and American contexts, notably among the younger cohorts. The majority of households still receive a broadcast signal, often bundled with internet service where cable+ subscription is the norm. Finally despite the increased mobility of television, the primary site of reception remains the household. Television 2.0 retains both its environmental and regulatory uses, although background viewing is no longer limited to live broadcast. As a leisure activity, viewing is imbricated into the routines of everyday life. Although much of television's flow is now viewer-centered, more focused, anticipatory fannish viewing may still require one's schedule to be organized around the original broadcast airdate. The role of the comfortable living room sofa and the large screen cannot be underestimated in facilitating social viewing with family or friends. Television 2.0 thus remains a domestic and social technology.

It is fair to say, however, that the institutional core of television has been destabilized as a result of its hybridization with the internet assemblage and the technological affordances of Web 2.0. The decentralized structure of the web, in combination with streaming and downloading platforms, holds out the potential for disruption by offering alternative modes of production and delivery. In 2010, Abigail De Kosnik argued that piracy via unauthorized downloading was the future of television; the TV 2.0 findings, however, suggest that the most measurable expansion of IPTV on a global scale is taking place in relation to subscription streaming services, led by Netflix. By making its original content available to all its subscribers simultaneously, it effectively is creating the first global television market. Subscription streaming is starting to have an impact on DVD sales, which will most certainly affect the television home after-sales market. In the United Kingdom for example, sales fell twenty-four percent in 2016 (Sweney, 2017).

In response, US networks are beginning to experiment with making available an entire season of episodes on their online platforms (e.g., NBC's short-lived *Aquarius*). At the time of writing, CBS aired the first two episodes of the new *Star Trek* series, *Discovery*. The rest of the series will be delivered weekly on its paid subscription streaming service, CBS All Access (whereas the entire series is being simultaneously broadcast in Canada on the Space network). Moreover, the American television industry continues to enter into international sale agreements to make its content more quickly available. For example, the 2016 Season 11 "revival" of *The X-Files* (Fox) and Season 7 of *The Walking Dead* (AMC) aired in the United States on Sunday nights and the next evening on Fox UK. Specialty networks such as HBO have set up international divisions (e.g., HBO Asia, to which Sashin, who resided in India, subscribed) to ensure their content is more accessible by authorized means. The same is true of the cable+ industry: in a "if you can't beat 'em, join 'em" move, these providers have programmed "channels" on their DVRs for streaming services such as Netflix, thereby bypassing the need for a media streaming device for the consumer. John Caldwell is correct when he states that "television as an institution has proven resilient in adapting to a series of fundamental economic, technical, and cultural changes" (2004, p. 43).

CONCLUSION

While the "Netflix factor" may be changing the assemblage's core, in some aspects, subscription streaming services are becoming more like networks. Netflix, in particular, has made a concerted effort to deal with its own "piracy" problem, by cracking down on the use of proxy servers to allow those outside the United States to gain access to its larger archive of content. It is also moving toward a DRM model by allowing some content to be viewed "offline" in some markets. By doing so Netflix is following the lead of the networks, which have been licensing downloads of episodes through iTunes for over ten years (albeit not with great success). The company is also following in the networks' footsteps by starting to release select series in weekly installments, one example being Season 2 of the CW's *Riverdale*. Finally, in September 2017 Netflix announced it had made a deal with the Canadian government to spend at least half a billion dollars over a five-year period to support the production and distribution of Canadian film and television series. Critics have suggested that it is a shrewd move to get tax payer subsidized content to stream globally while avoiding being directly regulated by the Canadian Radio and Television Commission (Doyle, 2017). In short, the core of the television assemblage is not being breached but rather reconfigured and reconsolidated as a BTV-IPTV model that is both national and global.

This book has attempted to demonstrate empirically the different kinds of engagement that viewers have with Television 2.0 in terms of household configuration, reception, and fan participation. Inside the household, services, screens, and devices are organized into two intra-assemblages: BTV and IPTV. Most users, however, are HTV assemblers, who have both a broadcast signal and internet, and who use a combination of a TV set, computer, laptop, or mobile device to view television programming. The data presented have shown that television is really not quite as site unspecific as it was assumed to become. Most households have a TV set in the living room for the reason that the size and location generally offer the most comfort for both individual and social viewing. Of course streamed and downloaded content can be viewed on a TV set using a memory stick or media streaming device. Computer monitors are also large enough to serve as subsitutes for TV sets. For those who do most of their viewing on laptops, domestic arrangements are as powerful a determinant as any desire to overtly reject BTV. Mobile devices in particular are increasingly likely to serve as a second screen for either distracted or fannish viewing.

Viewing is also multimodal, with viewers engaging in a combination of live, time-shifted, and/or online viewing. There can be no question that the most traditional mode, with the exception of news and sports programming, is on the decline. Still, live viewing has specific functions not easily substituted by other modes. First it is still the first choice for background flow, and is thus useful for distracted multitasking or to punctuate time in one's daily routine. Content still matters in this context, with a preference for genres with more formulaic narrative structures or reruns of familiar texts especially in syndication. Its second use involves the opposite mode of reception—focused and attentive. In the Canadian, American, and British contexts, the original broadcast is the best option for devoted, anticipatory fans, particularly those who are active in fan communities or desire the experience of ambient affiliation through the following of Twitter hashtags.

Time-shifting through home recording technology is an established BTV practice that remains popular for a number of reasons. First it is a practice of resistance, breaking with the broadcast schedule for leisure and fannish viewing that is not highly anticipatory. In addition to enabling one to organize television around one's schedule and lifestyle rather than the other way around, viewers record programming in order to skip commercials or fast-forward/rewind programming and/or to rewatch to gain further pleasure from the narrative. Second, some viewers record programming as a backup in case they are not able to watch live. Time-shifting is also used to resolve scheduling conflicts with the network schedule or with those of other members of the household. Moreover, a DVR can be used to create an archive of multiple episodes; these can be marathoned, either to catch up with a series which is no longer a priority or as a preferred mode of reception. DVR technology with its live pause feature blurs the boundary between live viewing and time-shifting.

DVD technology also enables some of the same catch-up and rewatch practices. Fans who purchase DVDs also need to be understood as collectors who desire a material media object, not just an ephemeral flow or stream. Yet as already noted, the future of this technology is uncertain. One participant raised some questions about that future and the implications for content producers and broadcasting networks:

> I mean, I am starting to see a day where DVDs are just going to be obsolete to me because everything I want is in the Cloud. So it's coming pretty quick. The more Wi-Fi coverage we get, the more likely it's going to be that I am going to stop having physical things and start paying for access rather than for shows. ... So I wonder sometimes how does the money get in the right hands? If I am paying Hulu $7.95 a month and I watch fifteen hours of *Stargate* and two hours of other stuff, does that much money go to *Stargate* or at least to Sci-Fi [network]? (Knitmeapony)

CONCLUSION

A niche market will almost certainly continue to exist as long as the industry continues to produce TV box sets. If not, then the DVD is destined to become what Williams (1977) calls a *residual* technology.

As with time-shifting, viewers stream content as both backup and catch-up, both for missed episodes and entire seasons and for series of which one has become a fan into or after the completion of its broadcast run. Such catch-up is closely linked to marathon viewing. In the case of original series produced by Netflix and other streaming services, anticipatory fannish viewing and marathon viewing merge. Unauthorized downloading remains the only viable option for two groups of anticipatory fans who do not have access to the original live broadcast: those outside Canada, the United States, and the United Kingdom who do not have any other means of timely access to American and British series; and those residing in these countries who have overtly rejected BTV by cutting the cord, never getting a cable+ subscription in the first place and/or not using the latter even though it is available in the household. While live viewing and unauthorized downloading appear to be polar opposites in their respective alignments with BTV and IPTV, they are really two sides of the same coin. Indeed the emergence of unauthorized live streaming, although today primarily associated with sporting events rather than serial television, further disrupts the BTV/IPTV boundary. In sum, choices around modes of viewing have less to do with technology and more to do with domestic and affective relations. As such Television 2.0 is as central to the household and the rhythms of everyday life as radio was after its domestication. If anything, Television 2.0 offers even more opportunities for viewers to develop a range of affective relationships with television texts, including those produced and distributed by subscription streaming services.

The findings I have presented also demonstrate that the rhizomatic linkages between participatory culture and television are far more extensive today than in the past. Hybridization with the internet meant that engagement and community making with other fans outside of one's existing "real life" social networks were no longer limited to the local fan club, reading/producing fanzines, sharing vids via a postal chain, or attending a convention. The exponential growth of fan-and network-administered websites, archives, and forums not only created a vibrant online fan culture around popular television in particular but made it visible and therefore more accessible to new fans. While participation has unquestionably increased as a result, I have argued that fan practices need to be dehomogenized and placed along a continuum.

Most fans are unlikely to do much more than to seek out information or to visit discussion forums to "check" their reactions but not post, practices that involve no or little involvement in fandom. Their invisibility to fan studies researchers has resulted in the overrepresentation of those who are actively involved in interpretative and creative communities.

Social media has brought about a number of changes to both viewing and participation. First by liking on Facebook and following on Twitter, a set of affective relationships are made visible to one's social networks. Second, the use of a second screen to follow hashtags while viewing on a television or computer screen creates a sense of ambient affiliation. Twitter also reconfigures producer–fan and actor–fan relations, offering the latter a more immediate sense of connection and the potential for direct communication with actors, celebrities, and producers even if that potential remains largely unrealized. Contemporary social media can thus be said to have a power similar to that of television, one that John Hartley describes as "located in dirt" (1992, p. 23): Dirty power "resides in the interfaces between individuals, in ambiguous boundaries" (Leach, cited in Hartley, p. 23). Television 2.0 disturbs divisions not only between centralized broadcast transmission and diffuse internet distribution but also between participation and non participation. The architectures and functionalities of platforms such as Facebook and Twitter complicate the participatory continuum by effectively blurring the distinction between reception, information seeking, and interactivity.

As for online fan communities, social media platforms do not enable their formation and maintenance in the same way as the older platforms, due to their emphasis on weak rather than strong social ties and/or character limits that make in-depth, sustained interaction difficult. As Rosalind Gill and Keith Grint (1995) remind us, however, the use of technology is never limited to its original design and can always be repurposed. Facebook and Twitter serve to extend personal and community relationships already established on the older platforms. Finally, just as primary texts are more spreadable as a result of downloading and streaming platforms, so are fan-created secondary texts. This second order of spreadability both disrupts the established gift economy and troubles the boundaries of fan creative communities.

As with all empirical studies, TV 2.0 is not without its limitations. First it was intended to map out a wide range of viewing and participatory practices and to capture larger-scale patterns. Future empirical work that drills down into some of these individual practices related to either domestic or affective relations to capture more of the nuances is in order. For example, research done on television and everyday life often uses diaries to ensure more accurate reporting of viewing practices and modes of reception. Of course there are also practices and platforms that were not addressed. For example, I did not discuss Television 2.0 in relation to material fandom, including fan conventions, activism, other forms of cultural production, or other sites and platforms where fans congregate such as Reddit or Tumblr.[1] Second, while the TV 2.0 sample is an international sample, it is not truly global. Additional research is needed with viewers and fans in South America, Asia, and Africa in particular. Although the survey and interview questions did not specify, the respondents and participants overwhelmingly focused on English-language series from the United States and to a lesser extent the United Kingdom. Therefore, studies that capture engagement with both national and international content would be useful. Finally, it is important to keep in mind that the conclusions outlined above cannot be generalized to a general television viewing population. The TV 2.0 respondents were predominately white, well-educated, and media-savvy. The BTV intra-assemblage may even be more entrenched among populations from different racial, ethnic, and socioeconomic backgrounds. Media accounts of IPTV often assume a universal access to broadband when there are many racialized, economically deprived neighborhoods and remote regions even in North America where internet access may not even be available in the household. Conversely, IPTV may be even more prevalent among affluent youth and in countries such as South Korea and Japan, where digital television and mobile technologies were developed first.

In conclusion, the tele-technological assemblage of radio-television has never been static and the current Television 2.0 iteration, from the institutional core to the nodes of participatory culture, will continue to shift as a result of technological change. Yet, as long as television remains a domestic technology organized around centralized production and hybrid transmission/distribution, and fans develop affective relationships with industry-produced serial narratives, it is unlikely that a major reassemblage, or a third media age for that matter, is on the horizon any time soon.

Note

1. I did collect interview data on Tumblr but it is not presented here. See Bury (2017) for a discussion on its use by fans.

References

Bury, R. (2017). Technology, fandom and community in the second media age. *Convergence: The International Journal of Research into New Media Technologies, 23*(6), 627–642. doi: 10.1177/1354856516648084

Caldwell, J. (2004). Convergence television: Aggregating form and repurposing content in the culture of conglomeration. In L. Spigel & J. Olsson (Eds.), *Television after TV: Essays on a medium in transition* (pp. 41–74). Durham, NC: Duke University Press.

De Kosnik, A. (2010). Piracy is the future of television. Convergence Culture Consortium. Retrieved from http://convergenceculture.org/research/c3-piracy_future_television-full.pdf

Doyle, J. (2017, September 28). The Netflix deal is a very sweet deal for Netflix, not Canada. *The Globe and Mail*. Retrieved from https://beta.theglobeandmail.com/opinion/the-netflix-deal-is-a-very-sweet-deal-for-netflix-not-canada/article36421246/?ref=http://www.theglobeandmail.com&

Gill, R., & Grint, K. (Eds.). (1995). *The gender-technology relation: Contemporary theory and research*. London: Taylor & Francis.

Hartley, J. (1992). *Tele-ology*. New York: Routledge.

Sweney, M. (2017, January 5). Film and TV streaming and downloads overtake DVD sales for first time. *The Guardian*. Retrieved from https://www.theguardian.com/media/2017/jan/05/film-and-tv-streaming-and-downloads-overtake-dvd-sales-for-first-time-netflix-amazon-uk?CMP=share_btn_tw

Williams, R. (1977). *Marxism and literature*. Oxford: Oxford University Press.

APPENDIX

Television 2.0 Survey Questions

(Note: The survey has been modified in form from the online version)

Section A—Demographic and General Television Viewing Information

1. Age (Exclusion Question)
2. On average, how often do you watch television programming on television, computer, and/or mobile device? (Exclusion question)
3. Country of Residence
4. Postal/Zip code
5. Gender
6. Race/Ethnicity
7. Occupation

Section B—Television Screen

1. Are you able to receive broadcast television via cable, satellite, and/or antenna? (If no, go to Section C)

2. What percentage of your viewing of television programming is "live," that is, at the time of scheduled broadcast?
3. What percentage of your "live" television viewing is unplanned and/or could be considered "channel surfing?"
4. Do you use a PVR/DVR/"plus box" to record programming?
5. Do you have VOD (video on demand) service from your satellite or cable provider that allows you to select and view programming at your convenience?
6. What percentage of your television viewing is "time-shifted" to be viewed at a time different than the scheduled broadcast?
7. How often do you network your computer or other device (e.g., iPod) to your television set to watch television programming?
8. How often do you watch commercially produced DVDs of television programming on a DVD player connected to your television set?
9. In the past year, what percentage of the total amount of time that you spend viewing television programming has taken place in front of a television set?

Section C—Computer Screen

In the past year, have you watched television programming on your desktop and/or laptop computer? (If no, go to Section D)

1. How often do you watch broadcast television on your computer? (Frequently, Sometimes, Occasionally, Almost Never).
2. How often do you use your computer's DVD player to watch television programming? (Same scale as 1)
3. How often do you watch television programming streamed from network websites (e.g., ABC.com) or cable/satellite provider websites (e.g., Comcast)? (Same scale as 1)
4. How often do you watch television programming streamed from third-party sites (e.g., Hulu, YouTube, Netflix)? (Same scale as 1)
5. How often do you download television programming from a third-party site (e.g., iTunes, Pirate Bay)? (Same scale as 1)
6. What percentage of the total amount of time that you spend viewing television programming takes place in front of a computer? (1–100%)
7. Why do you watch television on the computer?

☐ Programming not available through cable/satellite or direct TV service where I live.
☐ Programming is available but I prefer to watch it on the computer.
☐ Programming available on station/specialty cable network but I do not have a subscription.
☐ Programming available online before it is broadcast where I live.
☐ Don't have a television set.
☐ Other (Specify)

Section D—Mobile Television Screen. (Cell Phone, iPod, iPad)

1. In the past year, have you watched television programming on a mobile device such as a cell phone, iPod, or iPad? (If no, go to Section E)
2. Please indicate which service(s) you have used:
   - iTunes
   - Other: specify
3. What percentage of the total amount of time that you spend viewing television programming involves a mobile device?
4. Under what circumstances do you watch television programming on a mobile device?

   ☐ Traveling
   ☐ Commuting
   ☐ At school
   ☐ At work
   ☐ At home
   ☐ Other (Specify)

Section E—Multiple Modes of Viewing

This section provides a list of different types of television programming and asks you to identify the delivery platform(s) you use for each type of programming that you have watched in the past year. The distinctions between the categories may not be clear-cut. For example, *The Dog Whisperer* could be classified as Reality (nonvoting), Educational, or Lifestyle—it is your call! E2 gives you the option to comment further on your viewing experiences.

E1. In the past year, which modes of delivery have you used to view the following types of television programming?

- Don't watch
- Broadcast (Live or Recorded) Internet
- Internet (Streamed or Downloaded)
- DVD (commercially produced or burned)

Use the radio buttons to select one or more for each of the following:

a) Animated
b) Children and Youth
c) Comedy and Satire
d) Daytime Talk Show
e) Documentary
f) Drama
g) Educational/History
h) Late Night Talk Show (e.g., The Daily Show)
i) Lifestyle (e.g., cooking, home improvement)
j) Movies (made for TV)
k) News/Current Affairs
l) Reality (nonvoting)
m) Reality (voting)
n) Situational Comedy (sitcom)
o) Soap Opera
p) Sports

E2. Have your television viewing patterns changed in recent years because of time-shifting, downloading, streaming, and mobile technologies?

Section F—Fandom in the Web 2.0 Era

Screening Question: Do you consider yourself a fan based on following activities? (If no, proceed to the end of the survey)

- Having watched the majority of episodes of at least one season of a particular series.
- Staying abreast of news about the series through traditional and new media.
- Accessing and/or using additional web content such as webisodes, quizzes, or games.

- Discussing the series with other fans on discussion forums or social networking sites.
- Producing creative work (for e.g., fiction, art, music, and/or videos) based on at least one favorite series.

1) How many television series have you watched over the past year of which you are a fan? (You should include series that are no longer in production).
2) Please provide the title of one series of which you are a fan:
3) At what point did you become a committed viewer of this series?
4) How many seasons of this series have you watched ...
    a) in total?
    b) at the time of scheduled broadcast, on demand, and/or recorded on VCR/DVR/PVR?
    c) on commercially produced DVD (purchase, gift, or rental)?
    d) streamed on or downloaded from the internet (does not include webisodes, covered in F20)?
    on a mobile device?
5) Have any new episodes of this series aired in the past year?
6) Over the past year, how many new episodes of this series have you watched ...
    a) live on the original airdate?
    b) as a repeat broadcast?
    c) recorded on VCR/DVR/PVR?
    d) on demand?
    e) streamed on or downloaded from the internet?
7) How did you watch repeat episodes of this series?
    a) I did not watch any episodes more than once.
    b) Repeat broadcast
    c) On demand
    d) VCR/DVR/PVR/"plus box" recording
    e) Internet
    f) Mobile device
8) Have you influenced others to become a fan of this series?
9) Have others influenced you to become a fan of this series?
10) Do you have a different viewing pattern for another series of which you are a fan?
11) If yes, provide the title of this series. (Questions 12–18 are the same as 3–9)

12) At what point did you become a committed viewer of this series?
13) How many seasons have you watched?
14) Have any new episodes of this series aired in the past year?
15) Over the past year, how many new episodes of this series have you watched?
16) How did you watch repeat episodes of this series?
17) Have you influenced others to become a fan of this series?
18) Have others influenced you to become a fan of this series?
19) In the past year, how often have you visited websites, online forums, blogs, social networking sites (e.g., LiveJournal), content sharing sites (e.g., YouTube), and/or used other forms of social media (e.g., Facebook, Twitter) in relation to any of the series of which you are a fan? (Do not just consider the series you named in F2 and F11.)
20) Do you access additional content on network websites, such as webisodes, games, quizzes, etc.?
21) Do you visit online discussion forums set up by the networks who broadcast the series, or independent sites such as Television Without Pity?
22) Do you participate in the discussions?
23) Do you participate in any fan-based activities on LiveJournal and/or Dreamwidth?
24) Have you included any of the series in your list of Facebook "likes?"
25) Do you post any comments on Facebook fan pages?
26) Do you use Twitter or follow a Twitter feed about any of the series?
27) Do you blog or add comments to blogs about any of the series?
28) Do you read fan fiction written about any of the series?
29) Have you written any fan fiction and shared it with other fans?
30) Do you watch fan videos based on any of the series?
31) Where do you view fan vids?
32) Have you made your own fan vids and shared them with other fans?
33) Have you attended a fan convention of any kind?
34) Please indicate the names of up to five cons, single-series, and/or multi-series (e.g., WorldCon, Comicon) that you have attended in the past year.
35) How has your level of involvement as a fan changed and/or how has your participation in fan activities changed in recent years as a result of Web 2.0 content-sharing and social networking technologies?

REFERENCES

208 Radio Luxembourg. (2001, February 7). History of Radio Luxembourg and its English service. Retrieved from http://www.radioluxembourg.co.uk/?page_id=2

Abercombie, N., & Longhurst, B. (1998). *Audiences: A sociological theory of performance and imagination*. London: Sage Publications Ltd.

ACMA. (2015). Subscription video on demand in Australia 2015. Retrieved from: http://www.acma.gov.au/theACMA/engage-blogs/engage-blogs/Research-snapshots/Subscription-video-on-demand

Adorno, T. W. (1954). How to look at television. *Quarterly of Film, Radio and Television*, 8(3), 213–235.

Ang, I. (1989). *Watching Dallas: Soap opera and the melodramatic imagination* (D. Couling, Trans.). New York: Routledge.

Bacon-Smith, C. (1992). *Enterprising women: Television fandom and the creation of popular myth*. Philadelphia, Pa.: University of Pennsylvania Press.

Barker, C. (1999). *Television, globalization and cultural identities*. Philadelphia: Open University Press.

Baym, N. K. (2000). *Tune in, log on: Soaps, fandom, and online community*. Thousand Oaks, CA: Sage Publications.

Bennett, J. (2008). Television studies goes digital. *Cinema Journal*, 47(3), 158–165.

Bennett, J., & Strange, N. (Eds.). (2011). *Television as digital media*. Durham, NC: Duke University Press.

Bennett, L. (2014). Tracing Textual Poachers: Reflections on the development of fan studies and digital fandom. *Journal of Fandom Studies*, 2(1), 5–20. doi: 10.1386/jfs.2.1.5_1

Bergreen, L. (1980). *Look now, pay later: The rise of network broadcasting*. New York: Doubleday.
Boddy, W. (1990). *Fifties television: The industry and its critics*. Chicago: University of Illinois Press.
Booth, P. (2010). *Digital fandom: New media studies*. New York: Peter Lang Publishing.
Bothun, D., & Lieberman, M. (2011, October). Speed of life: Discovering behaviors and attitudes related to pirating content. PricewaterhouseCoopers LLP. Retrieved from: http://www.pwc.com/us/en/industry/entertainment-media/publications.html
Bouckley, H. (2016, October 3). From Marconi and the transistor radio to DAB: the history of radio in the UK. British Telecom. Retrieved from http://home.bt.com/tech-gadgets/from-marconi-and-the-transistor-radio-to-dab-the-history-of-radio-in-the-uk-11364015764901
Bourdieu, P. (1993). *The field of cultural production: Essays on art and literature*. New York: Columbia University Press.
Branston, G., & Stafford, R. (2010). *The Media Student's Book* (5th ed.). London: Routledge.
Britzman, D. (1995). Beyond innocent readings: Educational ethnography as a crisis of representation. In W. Pink & G. Noblit (Eds.), *Continuity and contradiction: The futures of the sociology of education* (pp. 133–156). Cresskill, N.J.: Hampton Press.
Bruns, A. (2006). Towards produsage: Futures for user-led content production. In F. Sudweeks, H. Hrachovec & C. Ess (Eds.), *Cultural Attitudes towards Communication and Technology* (pp. 275–284). Tartu, Estonia.
Buechler, S. M. (2013). Mass society theory. In D. A. Snow, D. della Porta, B. Klandermans & D. McAdam (Eds.), *The Wiley-Blackwell encyclopedia of social and political movements*. Hoboken, NJ: John Wiley & Sons.
Bury, R. (2005). *Cyberspaces of their own: Female fandoms online*. New York: Peter Lang Publishing.
Bury, R. (2008). Remotely embodied friendships in female fan communities. In S. Holland (Ed.), *Remote relationships in a small world* (pp. 174–198). New York: Peter Lang Publishing.
Bury, R. (2017). Technology, fandom and community in the second media age. *Convergence: The International Journal of Research into New Media Technologies, 23*(6), 627–642. doi: 10.1177/1354856516648084
Bury, R. (2018a). Television viewing and fan practice in an era of multiple screens. In J. Burgess, T. Poell & A. E. Marwick (Eds.), *Sage handbook of social media* (pp. 372–389). Thousand Oaks, CA: Sage Publications.
Bury, R. (2018b). "We're not there." Fans, fan studies and the participatory continuum. In M. A. Click & S. Scott (Eds.), *The Routledge companion to media fandom* (pp. 123–131). New York: Routledge.
Bury, R., & Li, J. (2015). Is it live or is it timeshifted, streamed or downloaded? Watching television in the era of multiple screens. *New Media & Society, 17*(4), 592–610. doi: 10.1177/1461444813508368
Busse, K., & Hellekson, K. (2006). Introduction. In K. Hellekson & K. Busse (Eds.), *Fan fiction and fan communities in the age of the Internet: New essays* (pp. 5–32). Jefferson, NC: McFarland & Co.
Butler, J. (1990). *Gender trouble: Feminism and the subversion of identity*. New York: Routledge.

Cable Europe. (2014). Cable facts and figures. Retrieved from http://www.cable-europe.eu/industry-data/

Caldwell, J. (2004). Convergence television: Aggregating form and repurposing content in the culture of conglomeration. In L. Spigel & J. Olsson (Eds.), *Television after TV: Essays on a medium in transition* (pp. 41–74). Durham NC: Duke University Press.

Carlat, L. (1998). "A cleanser for the mind": Marketing radio receivers for the American home, 1922–1932. In R. Horowitz & A. Mohun (Eds.), *His and hers: gender, consumption, and technology* (pp. 115–137). Charlottesville: University Press of Virginia.

CBC. (2015). Growing number of Canadians cutting traditional television, CBC research shows. Retrieved from http://www.cbc.ca/news/business/growing-number-of-canadians-cutting-traditional-television-cbc-research-shows-1.3139754

CCTA. (2016). History of cable. Retrieved from https://www.calcable.org/learn/history-of-cable/

CMS. (2015). Canadian media statistics. Retrieved from http://canmediasales.com/canada-101/canadian-media-stats/

CRTC. (2015). Communications monitoring report 2015. Retrieved from http://www.crtc.gc.ca/eng/publications/reports/policymonitoring/2015/cmrre.htm#ex

Coppa, F. (2006). A brief history of media fandom. In K. Hellekson & K. Busse (Eds.), *Fan fiction and fan communities in the age of the Internet* (pp. 41–59). Jefferson, NC: McFarland & Co.

D'heer, E., & Courtois, C. (2016). The changing dynamics of television consumption in the multimedia living room. *Convergence: The International Journal of Research into New Media Technologies, 22*(1), 3–17. doi: 10.1177/1354856514543451

Dawson, M. (2007). Little players, big shows: Format, narration, and style on television's new smaller screens. *Convergence: The International Journal of Research into New Media Technologies, 13*(3), 231–250. doi: 10.1177/1354856507079175

De Kosnik, A. (2010). Piracy is the future of television. Convergence Culture Consortium. Retrieved from http://convergenceculture.org/research/c3-piracy_future_television-full.pdf

Deleuze, G., & Guattari, F. (1987). *A thousand plateaus: Capitalism and schizophrenia* (B. Massumi, Trans.). Minneapolis: University of Minnesota Press.

Douglas, S. J. (1999). *Listening in: radio and the American imagination, from Amos 'n' Andy and Edward R. Murrow to Wolfman Jack and Howard Stern* (1st ed.). New York: Times Books.

Doyle, J. (2017, September 28). The Netflix deal is a very sweet deal for Netflix, not Canada. *The Globe and Mail*. Retrieved from https://beta.theglobeandmail.com/opinion/the-netflix-deal-is-a-very-sweet-deal-for-netflix-not-canada/article36421246/?ref=http://www.theglobeandmail.com&

Early Television Museum. (2017). Mechanical Television. Retrieved from http://www.earlytelevision.org/bell_labs.html

Evans, E., McDonald, P., Bae, J., Ray, S., & Santos, E. (2016). Universal ideals in local realities: Online viewing in South Korea, Brazil and India. *Convergence: The International Journal of Research into New Media Technologies.* Advance online publication. doi: 10.1177/1354856516641629

Fiske, J. (1987). *Television culture*. New York: Methuen & Co.

Fiske, J. (1989). Moments of television: Neither the text nor the audience. In E. Seiter, H. Borchers, G. Kreutzner & E. M. Warth (Eds.), *Remote control: Television, audiences, and cultural power* (pp. 56–78). New York: Routledge.

Frankel, D. (2007). DVD timeline: Looking back at the format's history. Retrieved from http://variety.com/2007/digital/features/dvd-timeline-1117963613/

Frankenberg, R. (1993). *White women, race matters: The social construction of whiteness*. Minneapolis: University of Minnesota Press.

Gallant, M. (2008). Using an ethnographic case study approach to identify socio-cultural discourse: A feminist post-structural view. *Education, Business and Society: Contemporary Middle Eastern issues, 1*(4), 244–254.

Gauntlett, D., & Hill, A. (1999). *TV living: Television, culture and everyday life*. London; New York: Routledge.

Geertz, C. (1973). Thick description: Toward an interpretive theory of culture. In *The interpretation of cultures: Selected essays* (pp. 3–30). New York: Basic Books.

Ghadialy, Z. (2006). Mobile TV technologies. Retrieved from http://www.3g4g.co.uk/Other/Tv/Presentations/mobile_tv_introduction.pdf

Gibbs, S. (2015, November 10). Betamax is dead, long live VHS. *The Guardian*. Retrieved from https://www.theguardian.com/technology/2015/nov/10/betamax-dead-long-live-vhs-sony-end-prodution

Gill, R., & Grint, K. (Eds.). (1995). *The gender-technology relation: Contemporary theory and research*. London: Taylor & Francis.

Gray, J. (2003). New audiences, new textualities: Anti-fans and non-fans. *International Journal of Cultural Studies, 6*(1), 64–81.

Gray, J., Sandvoss, C., & Harrington, C. L. (Eds.). (2007). *Fandom: Identities and communities in a mediated world*. New York: New York University Press.

Greenberg, J. M. (2008). *From Betamax to Blockbuster: Video stores and the invention of movies on video*. Cambridge, MA: MIT Press.

Greenfield, R. (2012). Forget cord-cutters: Cable companies should worry about cord-nevers, *The Atlantic*. Retrieved from http://www.thewire.com/technology/2012/08/forget-cord-cutters-cable-companies-should-worry-about-cord-nevers/55380/

Gripsrud, J. (2004). Broadcast television: The chances of its survival in the digital age. In L. Spigel & J. Olsson (Eds.), *Television after TV: Essays on a medium in transition* (pp. 210–223). Durham NC: Duke University Press.

Grossberg, L. (1992). Is there a fan in the house?: The affective sensibility of fandom. In L. A. Lewis (Ed.), *The adoring audience: Fan culture and popular media* (pp. 581–590). London; New York: Routledge.

Haraway, D. (1988). Situated knowledges: The science question in feminism and the privilege of partial perspective. *Feminist Studies, 14*(3), 575–599.

Harris, S. (2015, November 9). Cord-nevers could be bigger threat to TV than cord-cutters. CBC. Retrieved from http://www.cbc.ca/news/business/cord-nevers-cord-cutters-tv-1.3308072

Harris, S. (2016, November 16). $25 basic TV can't stop customers from cutting their cable in record numbers. CBC. Retrieved from http://www.cbc.ca/news/business/basic-tv-cord-cutting-cable-1.3847342

Hellekson, K., & Busse, K. (Eds.). (2006). *Fan fiction and fan communities in the age of the Internet: New essays*. Jefferson, NC: McFarland & Co.

Hills, M. (2002). *Fan cultures*. London; New York: Routledge.

Hilmes, M. (1993). Invisible men: Amos n Andy and the roots of broadcast discourse. *Critical Studies in Mass Communication, 10*(4), 301–321.

Holt, J., & Sanson, K. (Eds.). (2014). *Connected viewing: Selling, streaming, & sharing in the digital era*. New York: Routledge.

Horkheimer, M., & Adorno, T. W. (2002). The culture industry: Entertainment as mass deception (E. Jephcott, Trans.). In G. S. Noerr (Ed.), *Dialetic of enlightenment: Philosophical fragments* (pp. 94–136). Stanford CA: Stanford University Press.

Horrigan, J. B., & Duggan, M. (2015). Home broadband 2015. Pew Research Center. Retrieved from http://www.pewinternet.org/2015/12/22/2015/Home-Broadband-2015/

Ihde, D. (1990). *Technology and the lifeworld: From garden to earth*. Bloomington: Indiana University Press.

Ipsos. (2014). Changing TV habits. Retrieved from http://ipsos-na.com/news-polls/pressrelease.aspx?id=6433

Jackson, J. (2016, Thursday January 7). Netflix: From DVD rentals to the verge of world domination. *The Guardian*. Retrieved from https://www.theguardian.com/media/2016/jan/07/netflix-streaming-ces-global-tv-network

Jacobs, J. (2011). Television interrupted: Pollution or aesthetic? In J. Bennett & N. Strange (Eds.), *Television as digital media* (pp. 255–280). Durham, NC: Duke University Press.

Jenkins, H. (1992). *Textual poachers: Television fans & participatory culture*. New York: Routledge.

Jenkins, H. (2006). *Convergence culture: Where old and new media collide*. New York: New York University Press.

Jenkins, H. (2007). Afterword: The future of fandom. In J. Gray, C. Sandvoss & C. L. Harrington (Eds.), *Fandom: identities and communities in a mediated world* (pp. 357–364). New York: New York University Press.

Jenkins, H., Ford, S., & Green, J. (2013). *Spreadable Media: Creating value and meaning in a networked culture*. New York: New York University Press.

Jenkins, S., & Walter, D. (2015). Esther the Wonder Pig. Facebook status update. Retrieved from https://www.facebook.com/estherthewonderpig/

Jenner, M. (2016). Is this TVIV? On Netflix, TVIII and binge-watching *New Media & Society, 18*(2), 257–273. doi: 10.1177/1461444814541523

Kirby, S. L., Greaves, L., & Reid, C. (2006). *Experience research social change: Methods beyond the mainstream* (2nd ed.). Toronto: Broadview Press.

Kompare, D. (2006). Publishing flow: DVD box Sets and the reconception of television. *Television & New Media, 7*(4), 335–360. doi: 0.1177/1527476404270609

Kompare, D. (2009). The benefits of banality: Domestic syndication in the post-network era. In A. D. Lotz (Ed.), *Beyond prime time: Television programming in the post-network era* (pp. 55–74). New York: Routledge.

Lawler, R. (2010). One-third of US adults skip live TV. Retrieved from http://gigaom.com/video/one-third-of-us-adults-skip-live-tv-report/

Leaver, T. (2008). Watching *Battlestar Galactica* in Australia and the tyranny of digital distance. *Media International Australia, 126*(February), 145–154.

Lee, P., & Talbot, E. (2016). There's no place like phone: Consumer usage patterns in the era of peak smartphone. Deloitte LLP. Retrieved from http://www.deloitte.co.uk/mobileUK/assets/pdf/Deloitte-Mobile-Consumer-2016-There-is-no-place-like-phone.pdf

Lewis, M. (2015, April 13). Canadians scrapping cable packages in larger numbers. *Toronto Star*. Retrieved from https://www.thestar.com/business/2015/04/13/canadians-scrapping-cable-packages-in-larger-numbers-report.html?referrer=http%3A%2F%2Ft.co%2FgGbkX0KJL4

Loechner, J. (2015). Time shifted TV viewing is the default. Center for Media Research. Retrieved from http://www.mediapost.com/publications/article/247581/time-shifted-tv-is-the-default.html

Lotz, A. (2009). What is U.S. television now? *The Annals of the American Academy of Political and Social Science, 625*(September), 49–59. doi: 10.1177/0002716209338366

Lotz, A. (2014). *The television will be revolutionized* (2nd ed.). New York: New York University Press.

Lubken, D. (2008). Remembering the straw man: The travels and adventures of hypodermic. In D. W. Park & J. Pooley (Eds.), *The History of Media and Communication Research: Contested Memories* (pp. 19–42). New York: Peter Lang Publishing.

Lull, J. (1990). *Inside family viewing: Ethnographic research on television's audiences*. New York: Routledge.

MacNeil, K. (2016). Torrenting *Game of Thrones*: So wrong and yet so right. *Convergence: The International Journal of Research into new media technologies*. Advance online publication. doi: 10.1177/1354856516640713

Marcus, G., E., & Saka, E. (2006). Assemblage. *Theory, Culture & Society, 23*(2–3), 100–109.

Marling, W. H. (2006). *How "American" is globalization?* Baltimore: The John Hopkins University Press.

Marwick, A. E., & boyd, d. (2011). To see and be seen: Celebrity practice on Twitter. *Convergence: The International Journal of Research into New Media Technologies, 17*(2), 139–158. doi: 10.1177/1354856510394539

Matelski, M. (1995). Resilient radio. In E. C. Pease & E. E. Dennis (Eds.), *Radio: the forgotten medium* (pp. 5–14). New Brunswick, N.J., U.S.A.: Transaction Publishers.

McCarthy, A. (2001). *Ambient television: Visual culture and public space*. Durham, NC: Duke University Press.

McIntosh, P. (1993). White privilege and male privilege. In A. Minas (Ed.), *Gender basics: Feminist perspectives on women and men* (pp. 30–38). Belmont, Calif.: Wadsworth.

Mittell, J. (2006). Narrative complexity in contemporary American television. *The Velvet Light Trap, 58*(Fall), 29–40.

Morrison, S. (2014). The evolution of television delivery. Retrieved from http://leightronix.com/blog/the-evolution-of-television-delivery/

Nachman, G. (1998). *Raised on radio*. Los Angeles: University of California Press.

Netflix Media Center. (2017). About Netflix. Retrieved from https://media.netflix.com/en/about-netflix

Newman, M. Z. (2012). Free TV: File-sharing and the value of television. *Television & New Media 13*(6), 463–479. doi: 10.1177/1527476411421350

Nielsen Company. (2009, April). How DVRs are changing the television landscape. Retrieved from http://www.nielsen.com/us/en/insights/news/2009/how-dvrs-are-changing-the-television-landscape.html

Nielsen Company. (2016a). The Nielsen Total Audience Report: Q1 2016. Retrieved from http://www.nielsen.com/us/en/insights/reports/2016/the-total-audience-report-q1-2016.html

Nielsen Company. (2016b). The Nielsen Total Audience Report: Q4 2016. Retrieved from http://www.nielsen.com/us/en/insights/reports/2017/the-nielsen-total-audience-report-q4-2016.html

Nielsen Company. (2016c). TV season 2015–2016 in review. Retrieved from http://www.nielsenadfocus.cn/us/en/insights/news/2016/tv-season-2015-2016-in-review-the-biggest-social-tv-moments.html

Ofcom. (2015). The communications market report 2015. Retrieved from https://www.ofcom.org.uk/research-and-data/multi-sector-research/cmr/cmr15

Oliveira, M. (2015, February 18). 1 In 10 English Canadians no longer watch TV, just web video: Poll, *Huffington Post (Canada)* Retrieved from http://www.huffingtonpost.ca/2015/02/18/nearly-1-in-10-anglophone_n_6707736.html

ONdigital. (2002). ITV digital history. Retrieved from http://www.onhistory.co.uk/

Oztam, & Nielsen Company. (2016). Australian multi-screen report: Q 01 2016. Retrieved from http://www.thinktv.com.au/content_common/pg-reports.seo

Page, R. (2012). The linguistics of self-branding and micro-celebrity in Twitter: The role of hashtags. *Discourse & Communication, 6*(2), 181–201. doi: 10.1177/1750481312437441

Parks, L. (2004). Flexible microcasting: Gender, generation and television-internet convergence. In L. Spigel & J. Olsson (Eds.), *Television after TV: Essays on a medium in transition* (pp. 133–156). Durham NC: Duke University Press.

Pearson, R. (2011). Cult television as digital television's cutting edge. In J. Bennett & N. Strange (Eds.), *Television as digital media* (pp. 105–131). Durham, NC: Duke University Press.

Poster, M. (1995). *The second media age*. Cambridge, MA: Polity Press.

Poushter, J. (2016). Smartphone ownership and internet usage. Pew Research Center. Retrieved from http://www.pewglobal.org/2016/02/22/smartphone-ownership-and-internet-usage-continues-to climb-in-emerging-economies/

Press Association. (2005, October 11). A history of the license fee. *The Guardian*. Retrieved from https://www.theguardian.com/media/2005/oct/11/bbc.broadcasting1

Quigley, J. (1992). Videocassette recorders. In D. Ulloth (Ed.), *Communication technology: A survey* (pp. 161–166). New York: University Press of America.

Russo, J. L. (2009). User-penetrated content: Fan video in the age of convergence. *Cinema Journal, 48*(4), 125–130. doi: 10.1353/cj.0.0147

Schiesel, S. (2004). File sharing's new face. *New York Times*. Retrieved from http://www.nytimes.com/2004/02/12/technology/file-sharing-s-new-face.html?src=pm

Schonfeld, E. (2010). Live TV Is for old people: Time shifting and online make up nearly half of all viewing. Retrieved from http://techcrunch.com/2010/08/30/video-time-shifting-online-half/

Silverstone, R. (1994). *Television and everyday life*. New York: Routledge.

Smith, A. (2012). The 'rise' of the connected viewer. Pew Research Center. Retrieved from http://pewinternet.org/Reports/2012/Connected-viewers.aspx

Spigel, L. (1992). Installing the television set: Popular discourses on television and domestic space, 1948–1955. In L. Spigel & D. Mann (Eds.), *Private screenings: Television and the female consumer* (pp. 3–40). Minneapolis, MN: University of Minnesota Press.

Spigel, L. (2004). Introduction. In L. Spigel & J. Olsson (Eds.), *Television after TV: Essays on a medium in transition* (pp. 1–34). Durham NC: Duke University Press.

Steirer, G. (2014). The personal media collection in the era of connected viewing. In J. Holt & K. Sanson (Eds.), *Connected viewing: Selling, streaming, & sharing media in the digital era* (pp. 79–95). New York: Routledge.

Sterling, B. (1993). A short history of the Internet. Retrieved from https://w2.eff.org/Net_culture/internet_sterling.history.txt

Sterling, C. H., & Kittross, J. M. (2002). *Stay tuned: A history of American broadcasting* (3rd ed.). New Jersey; London: Lawrence Erlbaum Associates, Publishers.

Sweney, M. (2017, January 5). Film and TV streaming and downloads overtake DVD sales for first time. *The Guardian*. Retrieved from https://www.theguardian.com/media/2017/jan/05/film-and-tv-streaming-and-downloads-overtake-dvd-sales-for-first-time-netflix-amazon-uk?CMP=share_btn_tw

Television Bureau of Canada. (2014). Numeris media technology trends. Retrieved from http://www.tvb.ca/pages/BBM_MediaTechnologyTrends_htm

Theberge, P. (2005). Everyday fandom: Fan clubs, blogging, and the quotidian rhythms of the Internet. *Canadian Journal of Communication*, 30(4), 485–502.

Think TV. (2016). A quick look at Fall 2015. *The Quarterly*. Retrieved from http://www.thinktv.com.au/media/stats_&_graphs/library/tv_is_everywhere.pdf

Thomas, J. (2011). When digital was new: The advanced television technologies of the 1970s and the control of content. In J. Bennett & N. Strange (Eds.), *Television as digital media* (pp. 52–75). Durham, NC: Duke University Press.

Tomlinson, A. (1990). Home fixtures: Doing-it-yourself in a privatized world. In A. Tomlinson (Ed.), *Consumption, identity, and style: Marketing, meanings, and the packaging of pleasure* (pp. 40–51). London; New York: Routledge.

Turkle, S. (1995). *Life on the screen: Identity in the age of the Internet*. New York: Simon & Schuster.

Turner, G. (2011). Convergence and divergence: The international experience of digital television. In J. Bennett & N. Strange (Eds.), *Television as digital media* (pp. 31–51). Durham, NC: Duke University Press.

Uricchio, W. (2004). Television's next generation: Technology/interface/culture/flow. In L. Spigel & J. Olsson (Eds.), *Television after TV: Essays on a medium in transition* (pp. 163–182). Durham NC: Duke University Press.

REFERENCES

US Census Bureau. (1999). *Selected Communications Media 1920–1998*. (No. 1440). Retrieved from https://www.census.gov/en.html.

Van Der Sar, E. (2016). *Game of Thrones* most pirated TV show of 2016. *TorrentFreak*. Retrieved from https://torrentfreak.com/game-of-thrones-most-torrented-tv-show-of-2016-161226/

Vonderau, P. (2014). Beyond piracy: Understanding digital markets. In J. Holt & K. Sanson (Eds.), *Connected viewing: Selling, streaming, & sharing in the digital era* (pp. 100–123). New York: Routledge.

Webster, J. G., & Phalen, P. F. (2009). *Mass audience: Rediscovering the dominant model*. New York; London: Routledge.

Williams, R. (1975). *Television: Technology and cultural form*. New York: Schocken Books.

Williams, R. (1977). *Marxism and literature*. Oxford: Oxford University Press.

Wilson, S. (2016). In the living room: Second screens and TV audiences. *Television & New Media, 17*(2), 174–191. doi: 10.1177/1527476415593348

Wohn, Y., & Na, E.-K. (2011). Tweeting about TV: Sharing television viewing experiences via social media message streams. *First Monday, 16*(3). Retrieved from http://journals.uic.edu/ojs/index.php/fm/article/view/3368/2779

WRAL. (2014, July 14). History of WRAL digital. Retrieved from http://www.wral.com/history-of-wral-digital/1069461/

INDEX

A

ABC (American Broadcasting Company), 19, 47
   iTunes content, 27
Abercombie, N., 92
Adorno, T., 2–3
affect, 5, 73–74, 88–89
   anticipation and, 80–83
   of collectors, 85–86
   intensity, 77–80
   marathon viewing and, 86–89
   making a commitment to a series and, 74–77
   repeat viewing and, 84–85
Africa, 117
Aimee (study participant), 41, 63
Alias, 85
Amazing Race, 94, 101
Amazon
   subscription streaming services, 29, 89
ambient affiliation, 99, 114
AMC network, 83, 86

American Idol, 66, 87, 95
Amos 'n' Andy, 18, 19
analog era of television, 21–24
analysis of variance (ANOVA), 8, 59
Animal Planet channel, 60
Anna A. (study participant), 81
Anna B. (study participant), 48, 51, 58, 62, 68, 104, 106
Anne C. (study participant), 87, 95, 103–106
Anne E. (study participant), 41, 45–46, 56, 58
Annika (study participant), 48, 93, 94
anticipation, 80–83
anti-fans, 80
Apple. *see* specific devices
Apple TV, 27
Argentina, 10, 51
Asia, 10, 117
assemblage, tele-technological, 2–7
   television as, 20, 22, 24–26, 30–31, 111–113
audience continuum, 92
audience measurement, 2, 18, 30, 31, 56
Australia, 10, 49, 82

American series in, 21
cable subscriptions in, 29
cell phone and tablet ownership in, 28
channel surfing in, 59
DTT users in, 29, 49
HBO in, 79
IPTV in, 40, 48, 53
programming flow in, 21
streaming in, 30, 39
switch to digital in, 25
time-shifting in, 38
unauthorized downloading in, 89

B

Bachelor, The, 101
background viewing, 5, 44, 47, 49, 56–58, 111
Bacon-Smith, C., 22, 101
Bae, J., 83
Barker, C., 21
Barthes, R., 5
Battlestar Galactica, 78, 87
Baym, N., 24
bedroom, TV viewing in the, 62–63
Being Human, 80
Belgium, 10
Bella (study participant), 83
Bell satellite TV service, 41–42
Bennett, J., 2,
Bennett, L., 93, 95, 99
Bergreen, L., 17
Big Bang Theory, The, 46, 59, 66, 77
Biggest Loser, The, 65
binge viewing, 80, 86–89. see also marathon viewing
BitTorrent, 26, 30, 39, 81–83
Boddy, W., 19
Body of Proof, 79
Bones, 83, 95–96
Boomtown, 79
Booth, P., 24, 105
Bothun, D., 30
Bouckley, H., 19
Bourdieu, P., 79

bourgeois aesthetics, 79, 88, 94
Boxee, 27
Boyd, d., 96
Branston, G., 20
Brazil, 10, 47–48, 83
Breaking Bad, 65, 76, 86
British Broadcasting Company (BBC). see also *Doctor Who*
　DVD series, 26
　early radio, 16, 18–19
　early television broadcasting, 19
　iPlayer, 27, 29–30, 46
　radio, 16, 18–19
　streaming service, 60
British Broadcasting Company (BBC) America, 83, 86
Britzman, D., 9
broadcast TV (BTV)
　centralized transmission, 16, 27, 31, 111
　engagement, 40–46
　intra-assemblage, 15, 30–31, 113, 117
　live viewing of, 28–29, 38–40, 41, 43–45, 56, 80–81, 113–115
　loyalists, 40, 43–46, 53, 55, 69
Brooke (study participant), 41, 46, 66
Bruns, A., 7
Buechler, S. M., 2
Buffy (study participant), 42, 57, 66, 76, 78, 80, 87, 96, 99
Buffy the Vampire Slayer, 58, 77, 84
　fans talking about, 98
Bunny (study participant), 43–44, 50, 56–57, 75, 77, 80, 84
Burn Notice, 79
Bury, R., 24, 79, 91, 94, 97, 101, 104, 117n
Busse, K., 24, 105
Butler, J., 101
BTV. see broadcast TV

C

Cable Europe, 29
cable
　cord cutting and, 28–29, 47–48, 81, 115

INDEX

delivery services, 3, 6, 24, 26–27, 42
  history of, 20–22
  penetration, 22, 29
cable+ services, 28–29, 60, 111–112
  BTV loyalists and, 40–42
  DVRs and, 26, 42, 112
  hybrid assemblers and, 49–50, 52
  IPTV trendsetters and, 47–49
  VOD services and, 41, 45–46
Caldwell, J., 112
Californication, 83
Camden (study participant), 44, 50, 52, 63, 65, 84, 94, 99
Camelot, 78
Canada, 10, 113–114
  American series in, 21
  BTV loyalists in, 40, 53
  cable subscriptions in, 29
  channel surfing in, 59
  commercial radio in, 16
  cord cutters in, 51
  DVR ownership in, 26
  IPTV trendsetters in, 47
  Netflix expansion into, 27
  programming flow in, 22
  smart phone and tablet ownership in, 28
  streaming in, 39
  switch to digital in, 25
  time-shifting in, 38–39
  unauthorized downloading of unavailable programs in, 115
Canada's Next Top Model, 78
Canadian Radio and Television Commission (CRTC), 28, 113
Caprica, 78
Carlat, L., 16–17
Castle, 51, 66, 77, 80, 95
catch-up viewing, 44–45, 52, 76, 87, 89, 114
CBC (Canadian Broadcast Network), 29, 41, 60
CBS (Columbia Broadcasting System)
  early radio, 17
  early television broadcasting, 19–20
  WRAL affiliate, 24–25
CBS All Access, 112

CCTA (California Cable & Telecommunications Association), 21–22, 25
cellular telephony, 27. *see also* smart phones
centralized transmission and privatized reception, 16, 111
channel surfing, 23, 59–60
Chicago Code, 96
CMS (Canadian Media Sales), 29
CNN (Cable News Network), 22, 59
collectors, 85–86
Community, 46, 65, 78
community antenna TV (CATV), 20–21
communities, fan, 101–106
computers
  viewing on, 1, 3, 30–31, 37, 45–46, 48–50, 58, 62, 113
  connecting to TV, 38, 40, 50
  laptops, 45, 62, 67–68
convergence, 25, 31, 42
Coppa, F., 22, 24, 102
cord cutting, 28–29, 47–48, 81, 115
Corsac (study participant), 50, 65, 83
Courtney (study participant), 52, 68, 98
Courtois, C., 64
co-viewing, 65–66, 68–70, 97. *see also* social viewing
creative fanworks. *see* fan fiction; fan videos
Criminal Minds, 78, 96
CSI, 52, 77
CSI New York, 80
CTV network website, 51
Curb Your Enthusiasm, 79
CW network, 113

D

Daisy (study participant), 41, 43, 59, 74, 77, 79
Dallas, 21
Dawson, M., 28, 64
De Kosnik, A., 112
Deleuze, G., 5, 30
demographic snapshot of Television 2.0 study, 9–10

Dexter, 74, 76–78
D'heer, E., 64
digital convergence and divergence, 25–26
digital television. *see* digitalization; internet protocol TV; Television 2.0
digital terrestrial television (DTT), 24, 29, 37
   BTV loyalists and, 40, 46
   HTV assemblers and, 49
   IPTV trendsetters and, 48–49
   poor reception issues, 47
digital versatile discs (DVD), 26, 31, 39, 93, 112, 114–115
   background viewing and, 58
   box sets, 25–26, 58, 76, 82
   catch-up viewing and , 76, 87
   collectors, 85–86
   intra-assemblage relations and, 39
   IPTV trendsetters and, 47–48
   repeat viewing and, 84
digital video recorders (DVR), 29, 30–31, 114. *see also* time-shifting
   BTV loyalists and, 40–41, 43–44
   demographic differences and, 38–39
   household penetration, 26
   "live-pause" functionality, 45
   repeat viewing and, 84
   scheduling conflicts and, 66–67
   marathon viewing and, 87–88
digitalization, 15, 24–25
direct effects model of mass communication, 2
Discovery Channel, 47, 60
discussion boards, 95, 107
DISH TV, 41
Diva (study participant), 57, 95, 102
divergence, 25, 31
Doctor Who, 22, 31, 43, 51, 74
   fan anticipation and, 80, 83
   fan commitment to series and, 77–78
   fan produced content and, 106
   fans talking about, 97–98
Dollhouse, 98
domestication of technology, 17
domestic culture, 4–5
Douglas (study participant), 44, 62–63, 76, 84, 87, 94, 97

Douglas, S., 16–17
downloading, unauthorized, 51–52. *see also* piracy; peer-to-peer file sharing
Downton Abbey, 74
dramas
   radio, 18
   television. *see* specific series
Duggan, M., 29
DSL (dedicated subscriber line), 25, 111
DTT. *see* digital terrestrial television
DVD. *see* digital versatile discs
DVR. *see* digital video recorders
Dynasty, 21

E

Early Television Museum, 20
Ellen (study participant), 45, 63, 69, 87–88, 97
Elly (study participant), 43, 45, 76, 79, 85, 86–87, 92, 104
ER, 43, 68
Esther the Wonder Pig, 1
Europe, 10. *see also* Western Europe
Evans, E., 83
Event, The, 74, 94
everyday life, 55–56, 114
   families that view together in, 64–69
   leisure viewing in,, 59–64
   structuring of, 56–58
   television remaining fully imbricated with, 69–70
Eye-TV, 49–50
EZ Boards, 98

F

Facebook, 1, 7–8, 12, 107–108, 116
   community making and, 103–104
   fan produced content on, 106
   fans talking about television on, 99
   information seeking on, 94–95
Fallon, Jimmy, 99

families that view television together, 64–69
fan clubs, 22
fandom, 4, 8, 24, , 91–92, 100–101, 107–108
    collectors, 85–86
    communities and, 101–106
    gift economy and, 24, 105, 108
fan fiction, 4, 91–93, 101, 103–105
fanforum.com, 98
fans
    affective relations, 51, 73, 75, 82, 95, 107
    anticipation, 80–83
    information seeking and, 93–97, 116
    making a commitment to a series, 74–77
    marathon viewing and, 86–88, 89
    participatory continuum and, 92–93
    reaction/interpretation of, 12, 92, 97–100
    repeat viewing and, 84–85
    as textual poachers, 4, 91
fan studies, 2, 4, 91, 108n
fan videos, 24, 104–106
Farah (study participant), 45, 60, 66, 97
Federal Communication Commission (FCC), 17–18
feminization of radio, 17–18
Fillion, Nathan, 95
Firefly, 79, 86
Fiske, J., 4, 5, 21–22, 73, 75, 93, 97
flow
    broadcast, 21, 112
    international/global 7, 83
    viewer-centered, 61
Food Network, 57, 60
Ford, S., 6, 82, 105
Fox network, 22, 26
    Hulu and, 27
France, 21, 83
Frankel, D., 25–26
Frankenberg, R., 10
Frankfurt School, 2
Freda (study participant), 49, 51, 58, 67, 68, 103, 105
Freeview service, 42
Friday Night Lights, 51
Friends, 43, 68, 85

Fry, Stephen, 95
FX channel, 41

G

Gallant, M., 8
Game of Thrones, 30, 74, 79, 80, 82, 93
Gauntlett, D., 23, 57, 59
Geertz, C., 8
Gene (study participant), 49–50, 106
George F. (study participant), 48, 51, 65, 68, 102
Germany, 10, 48
    American series in, 21
    cable subscriptions in, 29
    switch to digital in, 25
Ghadialy, Z., 27
Gibbs, S., 23
gift economy, 24, 105, 108, 116
Gill, R., 116
Glau, Summer, 86
Glee, 65, 77
global television flows, 7
Glocal (study participant), 59, 75
Google, 94
    TV, 27, 50
Gray, J., 91
Greaves, L., 8
Green, J., 6, 82, 105
Greenfield, R., 28
Grey's Anatomy, 44, 65, 74, 77, 79, 85
Grint, K., 116
Gripsud, J., 16
Grossberg, L., 5, 80
GSM (study participant), 48, 51, 58, 64, 106
Guattari, F., 5, 30
Gundam Wing, 75

H

Hanson, Hart, 96
Haraway, D., 40
Harrington, C. L., 91

Harris, S., 28, 29
Harry's Law, 66
Hartley, J., 116
hashtags, use of, 99–100
Hastings, Reed, 26
Hawaii Five-0, 66, 77, 92
Headline News, 59
HBO (Home Box Office), 22, 30, 60, 79, 112
    fan anticipation and, 80, 82
HDTV (high definition TV), 25
Helen (study participant), 50, 85, 87
Helfer, Tricia, 78
Hellekson, K., 24, 105
Hercules, 84
Heresluck (study participant), 47–48, 51, 58, 62, 76, 82, 86, 102–104
HGTV network, 41, 57
Hill, A., 23, 57, 59
Hilmes, M., 18
History Channel, 60
HitFlix, 94
Holt, J., 2
Honeymooners, The, 23
Horkheimer, M., 2–3
Horrigan, J. B., 29
House, 56, 66, 79–80
household assemblers, 37, 113
House of Cards, 27
How I Met Your Mother, 77
Hub Entertainment Research, 29
Hulu, 27, 29, 47, 89, 114
    fan commitment to series and, 76
    IPTV trendsetters and, 49
    unauthorized downloading vs., 52
Hulu Plus, 29
hybrid TV (HTV), 6, 26, 30
    assemblers, 37, 40–45, 49–50–52
    as future of television, 53, 111–112

I

Idoru (study participant), 42, 56, 81
Ihde, D., 40
impatience economy, 83
India, 10
information and communication technologies (ICTs), 6, 24, 97, 101, 104, 107
information seeking, 93–97, 116
In Plain Sight, 79
internet protocol TV (IPTV)
    decentralized transmission, 31, 112
    intra-assemblage, 15, 113, 117
    rise of, 26–28, 30–31
    trendsetters, 40, 47–53
    social viewing and, 68–69
    subscription streaming and, 52, 112
    unauthorized downloading and, 51, 115
intra-assemblages, 30, 37–40, 113, 117. *see also* broadcast TV; internet protocol TV
In Treatment, 87
iPad, 28, 63
iPhone, 27, 64
iPlayer, BBC, 27, 29–30, 46
iPod, 28, 51
Ipsos, 30
IPTV. *see* internet protocol TV
Israel, 10
Italy, 21
iTunes, 26, 27, 39, 113

J

Jackson, J., 27
Jacobs, J., 67–68
Jake (study participant), 49, 56, 58, 59–60, 62, 65
Japan, 21
Jayne (study participant), 42, 57, 61, 96
Jeeves and Wooster, 95
Jenkins, H., 4, 6, 12, 24, 82, 105
    on digitalization impact, 25
    on fan anticipation, 80
    on marathon viewing, 87
    on participatory culture, 91, 100
    on process of becoming a fan, 75
    on repeat viewing, 84
    on textual poachers, 4, 91
Jenkins, S., 1

Jenner, M., 87
Jeopardy, 60, 66
Jersey Shore, The, 94
Joan (study participant), 45, 57, 63, 80–81, 85
Julianna (study participant), 44, 51, 86

K

Karen (study participant), 57–58, 69, 76, 78, 83, 95, 102–103
Kate (study participant), 60, 65
Katz, E., 2
Khal (study participant), 50, 59, 60, 64, 82, 93, 96, 100
Killing, The, 79, 82
Kim (study participant), 43, 46, 61
Kirby, S., 8
Kittross, J. M., 18
Knitmeapony (study participant), 49, 52, 57, 76, 78–79, 87, 95, 97, 99, 114
Kompare, D., 21, 23, 25–26

L

LA Law, 78
laptops, viewing on, 45, 62, 67–68
Lassie, 20
LA Times, 22
Lauchita (study participant), 41, 50–51, 74–75, 79, 82–83, 87
Law and Order, 57, 65, 78
Law and Order: SVU, 56
Lawler, R., 28
Leaver, T., 82
Lee, P., 28
Leverage, 84
Lewis, M., 28
Li, J., 8, 53n
Libby (study participant), 49, 96
Lieberman, M., 30
Lie to Me, 83
LimeSurvey, 7

Lisa (study participant), 63, 68, 74
listservs, 8, 24, 95, 98, 102
LiveJournal, 102–105
"live pause," 45
live viewing, 38–40, 41, 43–45, 59–60, 113–114
  as background, 47, 49, 56, 58, 111
  decline of, 28–29, 39, 81
  and fans, 80–81
living room, TV viewing in the, 20, 50–51, 62, 67–68, 112
Liz (study participant), 94
Loechner, J., 29
Longhurst, B., 92
Lost, 44, 74–75, 79, 81, 85
  fan produced content and, 106
  fans talking about, 98
Lost Girl, 77
Lotz, A., 2, 21
Louise (study participant), 49
Lull, J., 5, 11, 55–57, 65
LWR (study participant), 43, 94

M

M (study participant), 48, 51, 68, 83, 85
MacNeil, K., 82
Madeleine (study participant), 48
Mad Love, 77
Mad Men, 65, 74–76
  fan anticipation and, 81–83
  marathon viewing and, 86
Malawi, 10
Mantegna, Joe, 78, 96
marathon viewing, 24, 86–89, 114–115. see *also* binge viewing
Marcus, G. E., 5
Margaret (study participant), 41, 46, 103
Margene (study participant), 57–58, 67, 98
Marling, W., 19–20
Marwick, A. E., 96
Marxist theory, 2
Mary (study participant), 40–41, 45, 65, 67, 78, 84–85, 95–96

mass communication, direct effects model of, 2
Matelski, M., 17–19
Max (study participant), 41
McCarthy, A., 3
McDonald, P., 83
McIntosh, P., 10
media streaming devices, 27, 31, 38, 53, 112
Media Technology Monitor, 29
menu surfing, 60–61
Mike and Molly, 77
Mittell, J., 79, 94
mobile phones. *see* smart phones
modes of viewing. *see* viewing, modes of
modes of reception, 5
Moffat, Steven, 74
Momoa, Jason, 93
Monty Python, 98
Morrison, S., 21
Motorway Cops, 58
multichannel universe, 11, 22, 31, 47
multimodal viewing, 39, 113

N

Na, E., 100
Nachman, G., 18–19
narrative complexity, 79
National Geographic Channel, 60
NBC (National Broadcasting Corporation)
  early radio, 17
  early television broadcasting, 19
  Hulu and, 27
  Thursday night line-up, 61
NCIS, 52
Nem (study participant), 57–58, 61, 83, 98, 103–104
Netflix, 26–27, 29–30, 86, 89, 112–113
  in everyday life, 61, 63
  fan commitment to series and, 76–77
  intra-assemblage relations and, 39
  iPhone, iPod, and iPad apps, 27–28
  IPTV trendsetters and, 47, 50
  marathon viewing and, 86
  original series, 115
  unauthorized downloading *vs.*, 52
Netherlands, the, 10, 25
  cable subscriptions in, 29, 41
Newman, M., 30
New Zealand, 10, 82–83
  channel surfing in, 59
  DTT in, 49
  IPTV in, 40, 53
  programming flow in, 21
  streaming in, 39
  switch to digital in, 25
  time-shifting in, 38
  unauthorized downloading in, 89
Nielsen Company, 2, 18, 25–26
  on cable penetration, 28
  on decline in TV viewing, 29
  on fans talking about television, 99
  on mobile phone and tablet use, 28
Nikita, 44
Nintendo Wii, 27, 50
North America. *see* Canada; United States, the
Norway, 10, 83
Notesofwhimsey (study participant), 44, 56, 66, 67–68, 80, 104
"nowness," of television, 73
NVivo, 8

O

Ofcom, 26, 29–30
Office, The, 78
ONdigital, 25
online viewing. *see* computers, viewing on; downloading, unauthorized; streaming
oral culture, 97
O'Reilly, Tim, 6
Orphan Black, 74
Oscars, The, 97
over-the-air (OTA) delivery systems, 20–21, 42
Oztam, 28, 29

P

Page, R., 99–100
Parks, L., 6
Parks and Rec, 78
participatory continuum, 92–93, 115
participatory culture, 6, 8, 91, 100–101. *see also* fandom
Pearson, R., 22
peer-to-peer (P2P) file sharing, 26–27. *see also* BitTorrent
Penguin (study participant), 42, 44, 81
personal video recorders (PVR), 26. *see also* digital video recorders
Peter P. (study participant), 47–48, 83
Pew Research Center, 2, 28–29, 64, 99
Phalen, P. F., 18
Phillipe (study participant), 47, 78, 88, 100
piracy, 26–27, 82, 89, 112. *see also* downloading, unauthorized
popular cultural capital, 75
Poster, M., 2, 6, 31
postmodern theory, 5
Poushter, J., 28
Press Association, 16
PricewaterhouseCoopers, 30, 82
produsage, 7
Psych, 79
Pushing Daisies, 51

Q

QSR NVivo software, 8
Queer as Folk, 59, 76, 105
Quigley, J., 23

R

Radio Corporation of America (RCA)
    early television manufacturing, 19
    radio broadcasting, 16–17

radio era, 15–19, 55
Rain (study participant), 57, 60
Ray, S., 83
Real Housewives, 66, 101
reassemblage, 6, 22, 30–31, 117
reception, 4. *see also* viewing, modes of
    home as primary site, 3
    modes of, 5
    studies, 2, 4
Reddit, 117
Red Dwarf, 98
Reid, C., 8
remote control devices, 23, 31, 79
Rene (study participant), 46, 61, 68–69, 81, 83, 99, 106
repeat viewing, 24, 84–85, 87, 114
reruns, 57, 59, 75–76, 84, 114
residual technology, 115
Revan (study participant), 50, 68, 87–88, 102, 105
rhizomatic TV, 28–30, 111–112, 115
Riverdale, 113
Road Wars, 58
Robert (study participant), 65, 104
Roddenberry, Gene, 22
Roku, 27
Russo, J. L., 105
Ryan, Shawn, 96

S

Saka, E., 5
Sandvoss, C., 91
Sanson, K., 2
Santos, E., 83
Sarnoff, David, 16–17, 20
Sashin (study participant), 41, 112
satellite
    technology, 6, 22, 111
    TV services, 27, 40–42. *see also* cable+ services
Say Yes to the Dress, 57
Schiesel, S., 27
Schonfeld, E., 28

second screen, mobile device as, 53n, 64, 69, 97, 99, 113, 116
screens
  choice of, 67
  multiple, 1, 37–38
  size of, 46, 50–51, 63, 67–68, 112
Seinfeld, 79
Sentinel, The, 84
Serbia, 10
seriality, 31, 73
serials, radio, 18
series, television. *see* specific series
set-top boxes, 25, 31
Sex and the City, 59, 66, 80–81, 85
sharing economy, 24, 105
Sherlock, 74
SideReel.com, 52
Silverstone, R., 3–5, 17, 55
Simpsons, The, 77
sitcoms, 18, 66, 74. *see also* specific series
Six Feet Under, 79
Sky TV service, 30, 41–42, 83. *see also* satellite, TV services
Skype
  participant interviews by, 7
smart phones, 27–28, 48, 51, 63–64
Smith, A., 28, 64, 99
social media, 7, 8, 107–108, 115–116. *see also* Facebook; Twitter
  community making and, 101–104
  fan produced content on, 106
  fans talking about television on, 97–101
  information seeking using, 93–97
  participatory continuum and, 92–93
social uses typology, 55
social viewing, 66, 68, 112–113. *see also* co-viewing
Sons of Anarchy, 96
Sophie (study participant), 49, 52, 68, 84
Sopranos, The, 79, 81, 85
South America, 10, 88, 117
South Asia, 10
South Korea, 27

So You Think You Can Dance, 45, 65, 68–69, 96
  fans talking about, 100
Space network, 86, 112
Spigel, L., 1, 6, 20
Spiner, Brent, 96
spreadability of media, 6
  second order of, 105–106, 116
SPSS software, 8
Stafford, R., 20
Stargate, 114
Stargate Atlantis, 78, 81, 93, 95
Stargate Universe, 78
Star Trek (home video), 23
Star Trek (television series), 4, 22, 74, 85–86
  fans talking about, 98
Star Trek fandom, 24, 31
Starz network, 41
Steirer, G., 85
Sterling, C. H., 18
Stevie (study participant), 62, 85, 103, 104
Strange, N., 2
streaming, 7, 27, 29–30, 39–40, 45–46, 51–52, 69, 77, 84, 86, 112, 115. *see also* video on demand (VOD) services; media streaming devices
subscription streaming services. *see* Amazon; Hulu; Netflix; streaming
Supernatural, 57, 68, 80, 81
Survivor, 66, 77, 94
Sutter, Kurt, 96
Suzie (study participant), 61, 66–67, 75–76, 98, 101
Sweden
  unauthorized downloading in, 82
Sweney, M., 112
syndication, 21–22, 31, 84, 86, 114

T

Tabatha (study participant), 45, 79
tablets, Internet-enabled, 27–28, 64. *see also* iPad
Talbot, E., 28

INDEX

Tarsus (study participant), 42, 62, 102, 104
Tasha (study participant), 43–45, 58, 94
telcos, 25–27. *see also* cable+ services
tele-technological assemblage, 5, 117
tele-technological system, 3
television. *see also* broadcast TV; digital terrestrial television; hybrid TV; internet protocol TV; viewing, modes of
    analog era of, 21–24
    as assemblage, 2–7, 20, 22, 24–26, 30–31, 111–113
    as background, 56–58
    domestic culture and, 4–5
    as domestic technology, 55–56
    early broadcasting, 19–21
    early manufacturers of, 19
    as "medium in transition," 1–2
    over-the-air (OTA) delivery systems, 20–21
    radio as precursor to, 15–19
    rhizomatic, 28–30, 111–112
    social uses of, 55, 67–69
    syndicated, 21–22, 31, 84, 86, 114
Television 2.0, 6–7
    book outline, 10–12
    demographic snapshot, 9–10
    researching, 7–9
    survey questions, 119–124
Television Bureau of Canada, 26
Television without Pity, 98, 100
Telus TV, 44
textual poachers, 4, 91
Theberge, P., 22
Think TV, 26
30 Rock, 78
Thomas, J., 23, 25
time-shifting, 23, 29, 31, 38–39, 113–115
    family viewing and, 66–67
    as viewer-centred flow, 43–45, 61
TiVo, 26, 42. *see also* digital video recorders
TLC network, 57, 60
TNT network, 57
Tomlinson, A., 18, 55
Top Chef, 87

Top Chef Masters, 82
Torchwood, 78
TorrentFreak, 30
torrenting, 26, 30, 39, 81–83
True Blood, 74, 77, 80
Tumblr, 117
Turner, G., 9
Tutors, The, 74
Twin Peaks, 4, 85
Twitter, 7–8, 12, 107–108
    community making and, 102–103
    fans talking about television on, 99–100
    information seeking on, 94–97
    second screen and, 99, 116
Two and a Half Men, 60

U

United Kingdom, the, 10, 114, 117
    American series in, 21, 83, 112
    BTV loyalists in, 40, 53
    cable subscriptions in, 29, 41
    cell phone and tablet ownership in, 28
    DTT/Freeview users in, 42
    DVR ownership in, 26
    early television in, 19–21
    fan fiction in, 93
    financing of production and transmission costs in, 16
    programming flow in, 21
    radio era, 15–19
    streaming in, 29–30, 39
    switch to digital in, 25
    time spent viewing television in, 29
    unauthorized downloading of unavailable programs in, 115
United States, the, 10, 112–114, 117
    BTV loyalists in, 40, 53
    cable subscriptions in, 29, 41
    cell phone and tablet ownership in, 28
    commercial radio in, 16
    cord cutters in, 51
    DVR ownership in, 26

early television in, 19–21
fan fiction in, 93
first audience radio survey in, 18
IPTV trendsetters in, 47
OTA/DTT users in, 40, 42, 49
programming flow in, 22
radio era, 15–19
streaming in, 52
switch to digital in, 25
time-shifting in, 38
time spent viewing television in, 29
unauthorized downloading of unavailable programs in, 115
UPN network, 22
Uricchio, W., 22–23
USA Network, 79
US Census Bureau, 18, 20
user-circulated content, 7
user-produced content, 7

V

Vampire Diaries, The, 44, 46, 77
Van Der Sar, E., 30
Vera (study participant), 41, 59–60, 78, 95, 102, 104–105
Verbotene Liebe, 106
Veronica Mars, 79
videocassette recorders (VCR), 23–24, 31, 43
video on demand (VOD) services, 27, 29, 45–46, 112
  intra-assemblage relations and, 39
  marathon viewing and, 86
vids, 24, 104–106
viewing, modes of, 8, 11, 113–115
  live, 28–29, 38–41, 43–45, 56, 80–81
  online. *see* downloading, unauthorized; streaming
  time-shifted, 23, 29, 31, 38–39, 43–44, 66
Viggo, 41
Vimeo, 47, 105
Virginia (study participant), 42, 46
Vonderau, P., 82

W

Walking Dead, The, 86, 112
Walter, D., 1
WB network, 22
Webster, J. G., 18
Web 2.0 technologies, 1, 6, 106
Western Asia, 10
Western Europe, 10, 82. *see also* specific countries
  cable subscriptions in, 29
  cell phone and tablet ownership in, 28
  early television in, 19
  financing of production and transmission costs in, 10
  IPTV in, 40, 53
  programming flow in, 22
  streaming in, 39
  time-shifting in, 38–39
  unauthorized downloading in, 82, 88
West Wing, 79, 84
White Collar, 80
Wii, Nintendo, 27, 50
Will (study participant), 50, 60, 94, 99
William (study participant), 59, 62, 83–84, 86, 98, 100
Williams, R., 3, 15, 25
  on broadcast flow, 21
  on broadcast model, 16
  on origins of television, 19
  on television program genres, 21
Willow (study participant), 43, 64–65, 85, 98
Wilson, S., 64
Wire, The, 79, 87, 93
Wohn, Y., 100

X

Xbox/Xbox 360, 27, 50
Xena, 84
X-Files, The, 25, 74, 79, 85, 112
  community making and, 104
  fans talking about, 98

Y

Yahoo! Groups, 98, 102, 104
YouTube, 12, 27, 39, 47
    fan produced content on, 105–106

Z

Zee (study participant), 43, 67, 80, 105

General Editor: *Steve Jones*

Digital Formations is the best source for critical, well-written books about digital technologies and modern life. Books in the series break new ground by emphasizing multiple methodological and theoretical approaches to deeply probe the formation and reformation of lived experience as it is refracted through digital interaction. Each volume in **Digital Formations** pushes forward our understanding of the intersections, and corresponding implications, between digital technologies and everyday life. The series examines broad issues in realms such as digital culture, electronic commerce, law, politics and governance, gender, the Internet, race, art, health and medicine, and education. The series emphasizes critical studies in the context of emergent and existing digital technologies.

Other recent titles include:

Felicia Wu Song
    Virtual Communities: Bowling Alone, Online Together
Edited by Sharon Kleinman
    The Culture of Efficiency: Technology in Everyday Life
Edward Lee Lamoureux, Steven L. Baron, & Claire Stewart
    Intellectual Property Law and Interactive Media: Free for a Fee
Edited by Adrienne Russell & Nabil Echchaibi
    International Blogging: Identity, Politics and Networked Publics
Edited by Don Heider
    Living Virtually: Researching New Worlds

Edited by Judith Burnett, Peter Senker & Kathy Walker
    The Myths of Technology: Innovation and Inequality
Edited by Knut Lundby
    Digital Storytelling, Mediatized Stories: Self-representations in New Media
Theresa M. Senft
    Camgirls: Celebrity and Community in the Age of Social Networks
Edited by Chris Paterson & David Domingo
    Making Online News: The Ethnography of New Media Production

To order other books in this series please contact our Customer Service Department:
(800) 770-LANG (within the US)
(212) 647-7706 (outside the US)
(212) 647-7707 FAX

To find out more about the series or browse a full list of titles, please visit our website:
WWW.PETERLANG.COM